IMAGES
of America

THE PACIFIC GROVE MUSEUM OF NATURAL HISTORY

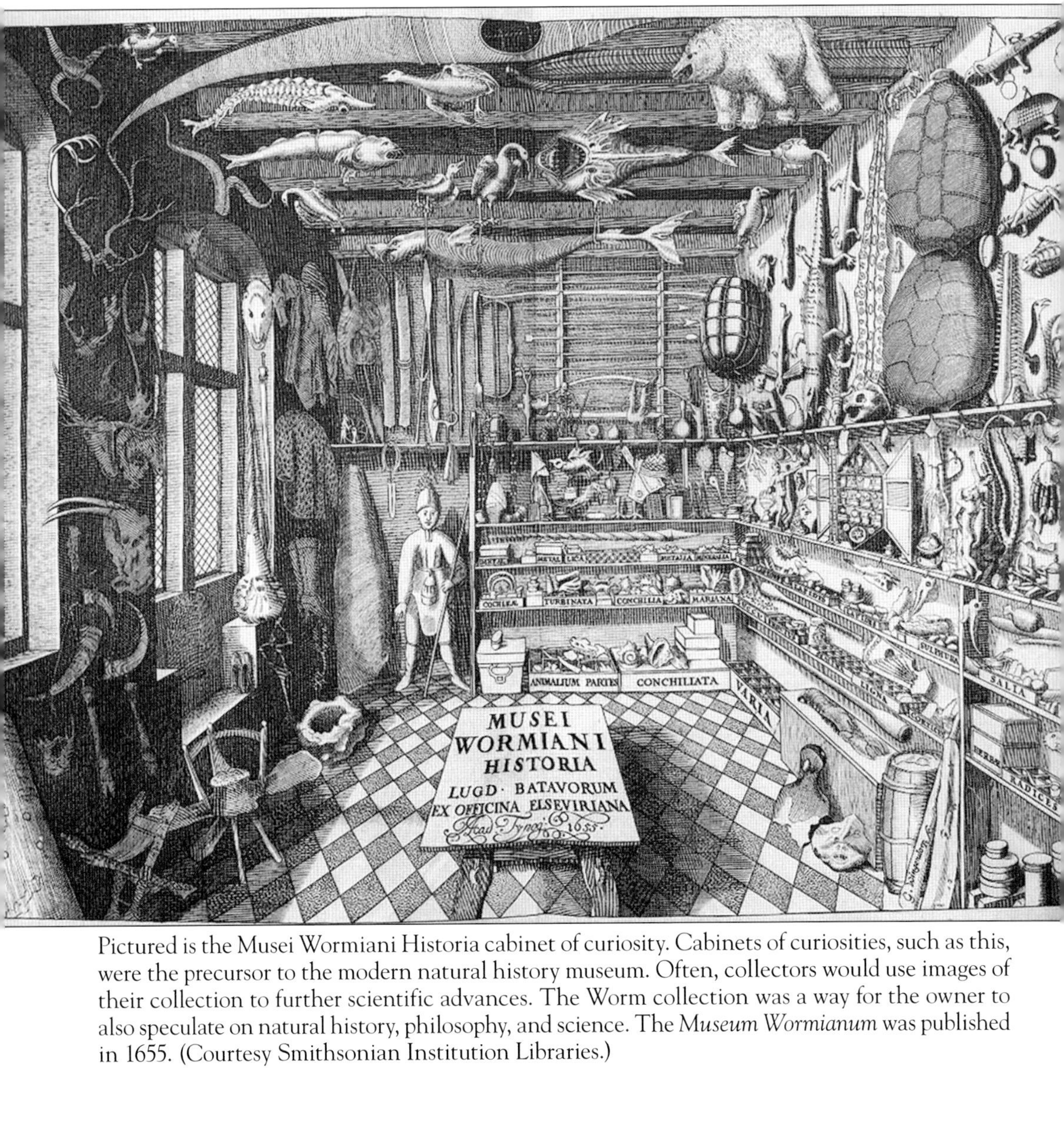

Pictured is the Musei Wormiani Historia cabinet of curiosity. Cabinets of curiosities, such as this, were the precursor to the modern natural history museum. Often, collectors would use images of their collection to further scientific advances. The Worm collection was a way for the owner to also speculate on natural history, philosophy, and science. The *Museum Wormianum* was published in 1655. (Courtesy Smithsonian Institution Libraries.)

ON THE COVER: This image from the C.K. Tuttle collection, dated around 1900, shows the original octagon-shaped Pacific Grove Museum. The view, seen from Central Avenue, formerly known as Grove Street, offers a glimpse of Forest Avenue. Men are also visible in the photograph as they play croquet nearby. The Chautauqua Literary and Scientific Circle used the structure during their two-week summer retreats. In 1932, it was replaced by the current museum building and has grown over the years. (Courtesy Pacific Grove Museum of Natural History.)

IMAGES
of America

The Pacific Grove Museum of Natural History

Patrick Whitehurst

ARCADIA
PUBLISHING

ISBN 978-1-4671-2833-9

Published by Arcadia Publishing
Charleston, South Carolina

Printed in the United States of America

Library of Congress Control Number: 2017952984

For all general information, please contact Arcadia Publishing:
Telephone 843-853-2070
Fax 843-853-0044
E-mail sales@arcadiapublishing.com
For customer service and orders:
Toll-Free 1-888-313-2665

Visit us on the Internet at www.arcadiapublishing.com

For Natalia Molina, who makes me work harder every day. Thank you.

Contents

ACKNOWLEDGMENTS

When it comes to the Pacific Grove Museum of Natural History, there is no such thing as loneliness. Staff work closely with community supporters, volunteers, and experts daily. When it comes to the book you now hold in your hands, the effort would not have been possible were it not for the assistance of a great host of natural history enthusiasts. In this instance, we focused that enthusiasm onto the natural history of the museum itself. While the museum prides itself on education and science, detailing histories both scientifically and historically significant, a record of the museum's growth was harder to come by. As it should, the museum has focused on the outside world's story more than its own story. That story can now be told, however, thanks to the community's help. Special thanks to museum executive director Jeanette Kihs, who worked tirelessly to create a museum the public can explore, learn from, and celebrate in. Her help and enthusiasm was invaluable in completing this project. Most days, curator Paul Van de Carr and collections registrar Nate King can be found at the museum moving collection items, up on ladders, or down in the basement. Their jobs take them to every nook and cranny of the historic landmark. Fortunately, it also provides two historical experts. I owe them a debt of gratitude for the knowledge they provided while fleshing out the chapters of this book. Thanks also to Mary Dainton, the museum's longest-running volunteer; curator emeritus Vern Yadon; Paul Finnegan; artist Larry Foster; Angel Hisnanick and Lindsey L. Givens with Arcadia Publishing; Don Kohrs; Dixie Layne with the Heritage Society of Pacific Grove; Diana Godwin with the City of Pacific Grove; museum docent Bill Wolfe; Dennis Copeland with the Monterey Public Library; and a host of others whose contributions made this book possible. Thank you all for making this trip into history a memorable one.

Introduction

When it comes to natural history, the emphasis of study often resides in observing subjects in their own environment and documenting those findings. In nature, that study can take the form of geology, animal, plant life, and include the study of fungi and humans. Since Ancient Greece, naturalists have documented their findings from the natural world, be it through drawings, logs, journals, and books with the hope to further understand the planet on which we all live. Through observation, they document how vegetation interacts with a particular environment, how it thrives, or how it is used by the animals around it. A naturalist often focuses on the relationship between one thing and another, the origin of a subject, and its contribution to the ecosystem. These days, findings from the natural world are still documented in the traditional sense, but also on the World Wide Web. Internet sites such as iNaturalist offer up-to-the-minute observations from naturalists scattered across the globe. Discovering the natural world around us has never been easier.

Collecting specimens of the natural world became something of an art form during the Renaissance with "cabinets of curiosities." Often referred to as "Theaters of the World," these cabinets are considered the building blocks of modern museums. A cabinet could describe a room full of marvels, from stuffed animals and relics, to rocks and fossils, many of which had yet to be categorized by science.

By the 19th century, natural history museums began to pop up around the world. These museums were used to preserve large collections for further study. They were also a way to educate the public on the wonders found in nature and provide a one-stop source of research for scientists in a field of study. It was not uncommon for museums to host more scientists than visitors. Some museums feature the natural history of the world, and feature skeletons of dinosaurs, a library of plant specimens, and much more. Others take a narrower focus and zero in on the natural history of a particular subject or locale.

Between the ocean-kissed beach at Lovers Point and the thriving hub of downtown Pacific Grove sits one of the oldest such institutions on the Central Coast. The Pacific Grove Museum of Natural History has been a part of the California coastal community since 1883. First found in the area where Jewell Park sits today, the museum began humbly in a small octagon-shaped building as something of a storehouse for members of the Chautauqua Literary and Scientific Circle after they ran out of room elsewhere. Many of the original collection items were gathered in and around the Monterey Bay coastal community. Items such as plant specimens were carefully preserved for later study by its members, a group that included both collectors and scientists.

Two such scientists, Prof. Josiah Keep, and Mary Norton, a botanist and geologist, respectively, are credited with the creation of the first Pacific Grove Museum. Keep himself would hold lectures and nature walks to the tide pools of Pacific Grove, where he amassed a large collection of shells now in the museum collection. Norton, meanwhile, operated with others to collect and mount seaweed specimens, as well as collections of wildflowers and native plants. The two worked with the local Pacific Improvement Company (PIC) to acquire a modest structure to keep materials collected in the Monterey Peninsula. The PIC was a subsidiary of the Southern Pacific Railway. The company built the Del Monte Hotel and the El Carmelo after they purchased the property from David Jacks in 1880. With these collections nestled comfortably under one roof, recording and documenting the items was made far easier. That collection continued to grow. By 1889, the group formed a museum association that included 15 charter members. Soon, that small structure was filled with more than 12,000 specimens from the natural world. It became clear more room would be needed. This need led to a sizable land donation by the PIC in 1902 that consisted of half a block just beneath Lighthouse Avenue, which today is prime Pacific Grove real estate. With the land in place, museum association members began raising the money needed to put a new, fireproof building on the land. By 1913, a new building had yet to appear, however, and the museum was offered to the City of Pacific Grove. The issue of ownership was decided four years later, with the city taking control of the museum in 1917. With this decided, plans for a larger museum were drawn up, but the building would not be completed until 1932, thanks to a $5,000 legacy donation from New York resident Dr. Anna Lukens, followed by a donation from Pacific Grove resident Lucie Chase that allowed construction on the new building to officially begin. A penthouse suite added in 1938 became the home for curator Teresa Lloyd and her young daughter, Myfanwy. The cost to add the suite, at that time, was just over $3,000. Construction costs changed a bit by the 1980s, when the museum expanded at a cost of $329,000. In 1983, the new wing was built, which included a basement for storage of collection items, handicap access, and public restrooms. It was at this time that Sandy the Whale arrived in Pacific Grove. This life-sized 40-foot replica of a gray whale played a part in paying for the new wing. Sandy the Whale, designed by artist Larry Foster, visited the museum during an exhibit on gray whales. Though she was originally only a temporary resident on a three-month "rental," the joy Sandy brought to residents led to an effort to purchase her outright. This attention helped raise money for the new wing as well, and both were quickly paid for through a combination of grants and donations. The new addition allowed museum staff the room to grow and expand their collection, not to mention grow the size of temporary exhibits, such as the large Dinosaurs Alive! exhibit that premiered in 1989. Families have found a treasure trove of educational material within the museum's walls, from samples of native plant life preserved and flattened in bound volumes, to shell collections and taxidermy that illustrate the complex and diverse ecosystem of the California Central Coast. This combined with lectures, special events, and workshops have created a unique niche in the area's educational landscape. Pacific Grove, and the entire Central Coast, just would not be the same without the natural history museum.

One

A City Is Born

Pacific Grove, originally a Methodist retreat founded in 1875, was once surrounded by a network of fences in an effort to preserve its virtue. That fence and gate eventually came crashing down around 1890, however, thanks to an ax owned by Judge Benjamin Langford. Langford used the ax to express his frustration at having to retrieve a key to unlock the gate to get to his home. After the gate fell, the fence was never rebuilt. An abundant population of breezy pine trees, coast live oak, cypress trees, and numerous wildflowers and native plants made the retreat ideal for camping. The city soon became a hub of growth, one that included the Chinese Fishing Village, which preceded the city by about 20 years, located along the area where Hopkins Marine Station can be found today. While fire destroyed that village in 1906, its memory continues. Descendants of the village still live in the area. Other churches, besides the Methodist retreat, also appeared in the mid-to-late 1880s, thanks to donations from the Pacific Improvement Company and landowner David Jacks. The Chautauqua Literary and Scientific Circle, which got its start at Lake Chautauqua in New York in 1874 as a Methodist training camp for Sunday school teachers, blossomed into a national movement and led to the first Pacific Grove "daughter" Chautauqua camp in 1879. While the first camps were housed in Retreat tents, Chautauqua organizers soon saw a need to grow their accommodations. This led them to the Old Chapel, a building also known as Boy Scout Hall. It is now more commonly known as Chautauqua Hall. The society soon recognized a need to store their collection of scientific items for further research. These items, collected from the shoreline and inland locations, would soon become the first collection items to be featured in the Pacific Grove Museum in 1881. They would not be the last.

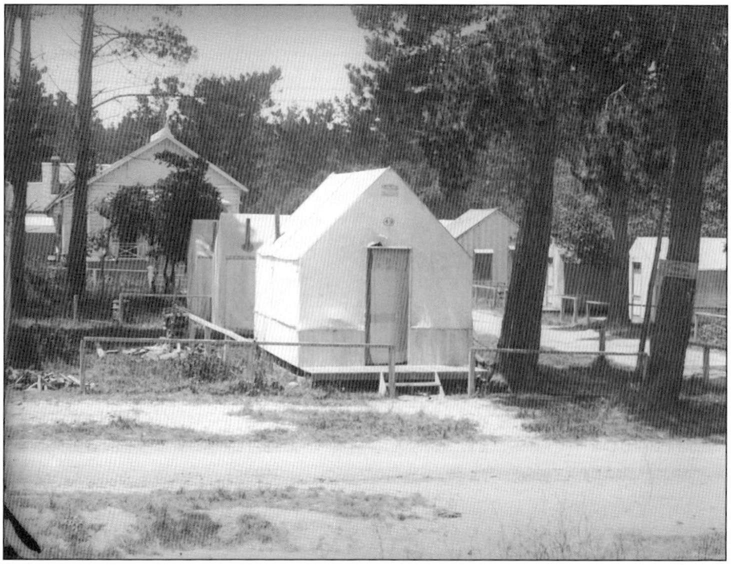

This image from the C.K. Tuttle collection, captured around 1900, depicts a Chautauqua Literary and Scientific Circle tent along Seventeenth Street, in what was then known as the Retreat section. The circle held a two-week assembly in Pacific Grove every summer, which featured lessons, exhibits, lectures, picnics, and concerts. (Courtesy Pacific Grove Museum of Natural History.)

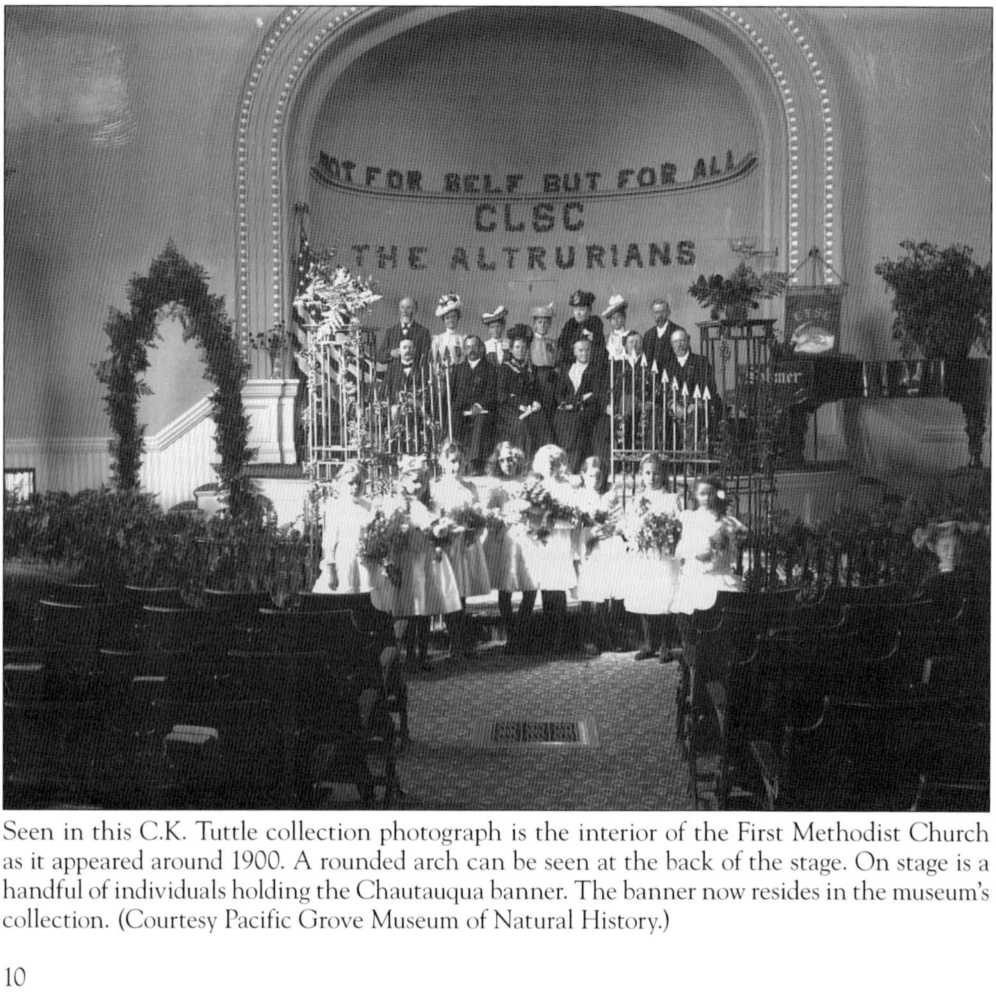

Seen in this C.K. Tuttle collection photograph is the interior of the First Methodist Church as it appeared around 1900. A rounded arch can be seen at the back of the stage. On stage is a handful of individuals holding the Chautauqua banner. The banner now resides in the museum's collection. (Courtesy Pacific Grove Museum of Natural History.)

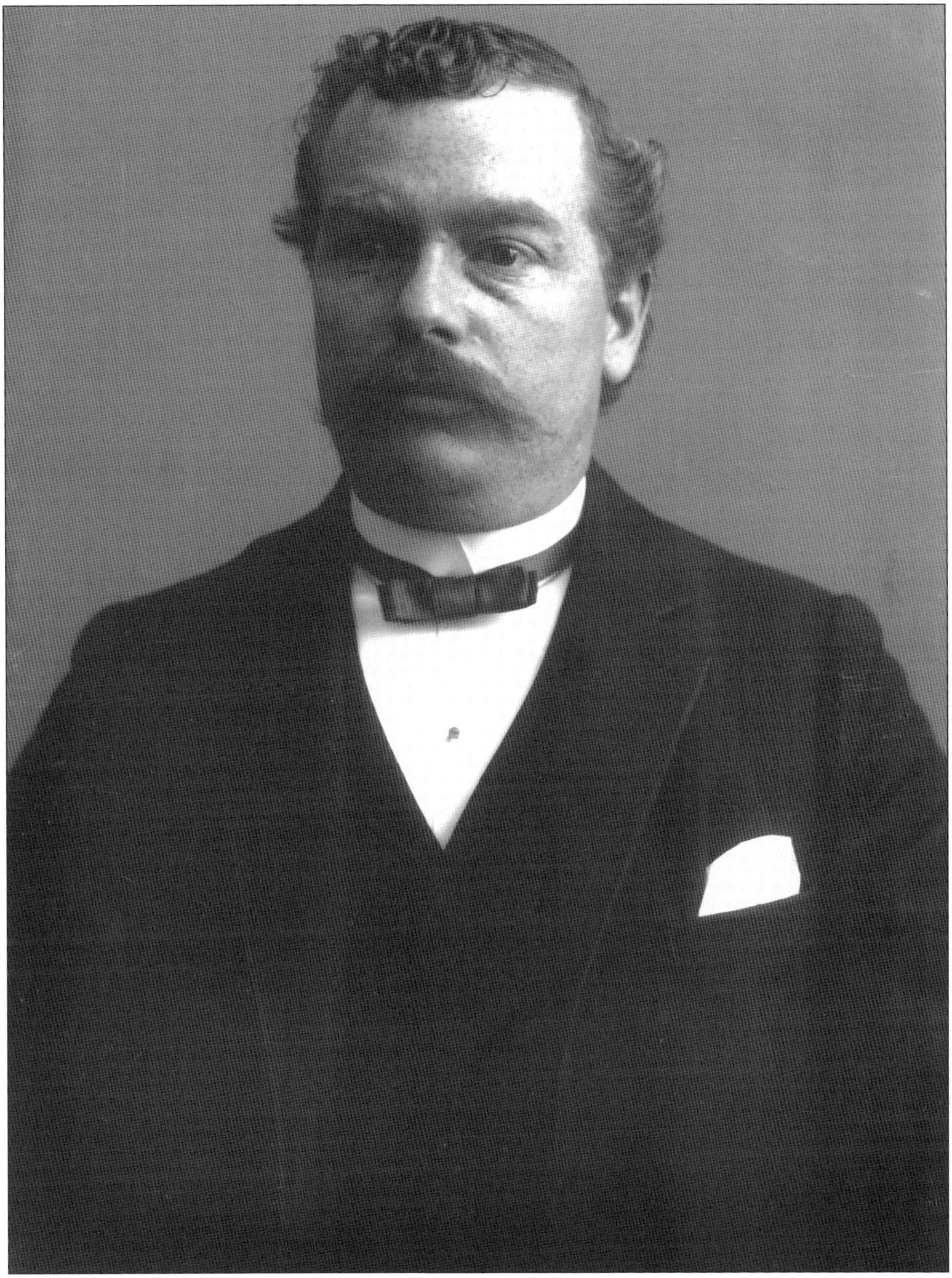

This is a 1905 portrait of photographer and Pacific Grove business owner C.K. Tuttle. Here, Tuttle is pictured with a walrus mustache, wearing a starched collar and bowtie. Besides his vast collection of photographs, now a part of the museum collection, Tuttle was also a well-known local pharmacist in Pacific Grove. Members of his family were visible in many of the photographs he took over the years. His collection of photographs is currently kept in the binders located in the museum's library. Besides Tuttle's photographs, the binders contain the work of other photographers from the area as well. (Courtesy Pacific Grove Museum of Natural History.)

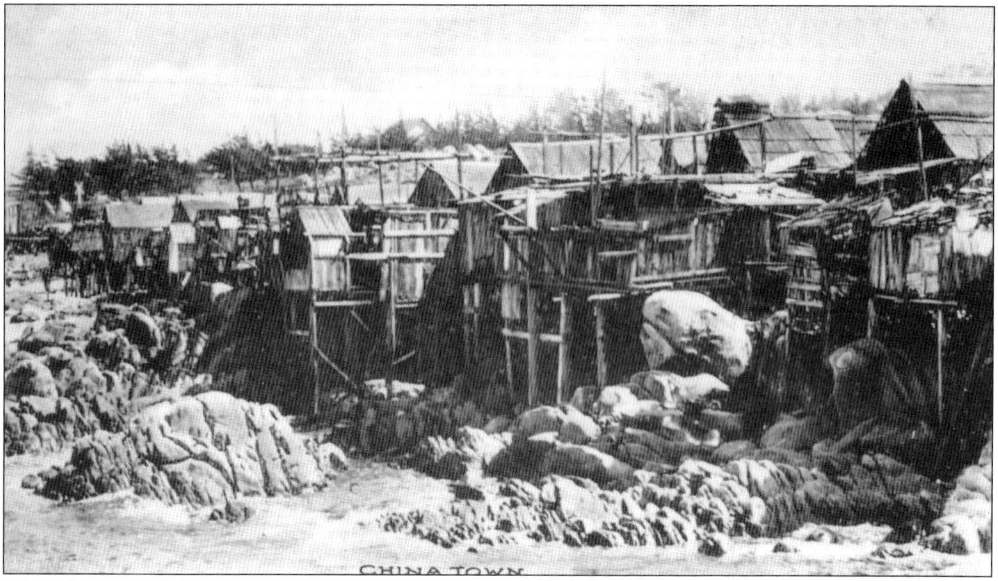

This photograph by C.W.J. Johnson from 1900 shows the Chinese Fishing Village, also known as China Town, a name visible at the bottom of the image. The wooden structures that made up this large community were located on the rocky shoreline near where the Monterey Bay Aquarium can be found today. A railroad crossing sign is also visible in the upper left-hand corner. (Courtesy Pacific Grove Museum of Natural History.)

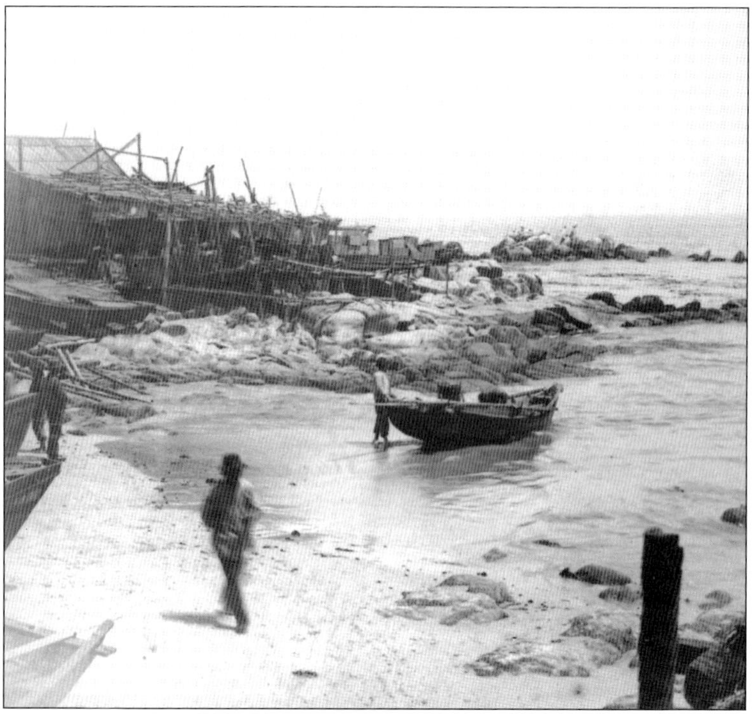

A boat is being launched in this photograph from 1900 taken at the site of the Chinese Fishing Village in Pacific Grove. The boats often contained baskets and other gear for fishing. The museum offers an exhibit still on display that offers the history of the Chinese community. The village, however, was destroyed by a mysterious fire six years after this image was captured. (Courtesy Pacific Grove Museum of Natural History.)

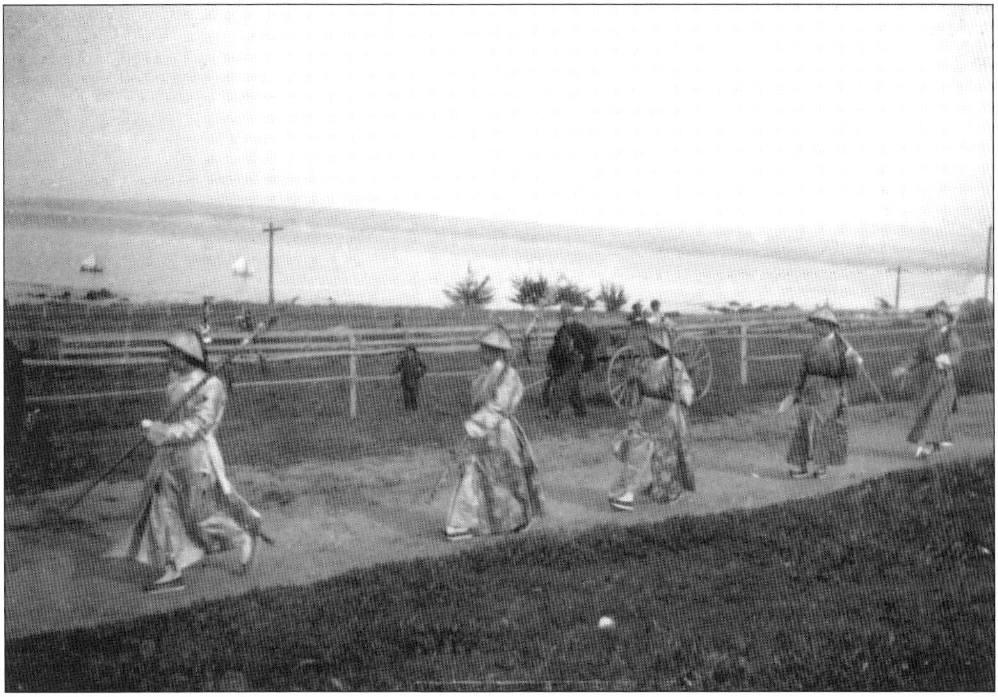

A Chinese procession moves along Ocean View Boulevard in this photograph from 1905. The procession is pictured between the Chinese Fishing Village (often referred to as China Town) and Lovers Point. Five men in silk robes carry embellished sticks in the photograph, possibly for the first Feast of Lanterns event, for a funeral procession. (Courtesy Pacific Grove Museum of Natural History.)

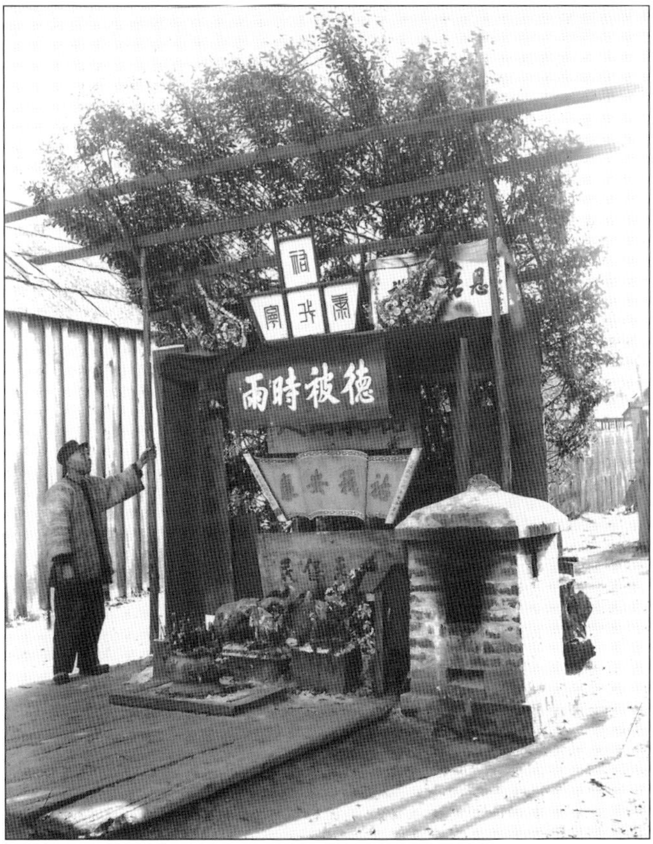

This 1900 photograph shows a man looking at Joss House, a temple altar located at the Chinese Fishing Village found until the year 1906 along the Pacific Grove shoreline at China Point, now the current site of the Hopkins Marine Station. (Courtesy Pacific Grove Museum of Natural History.)

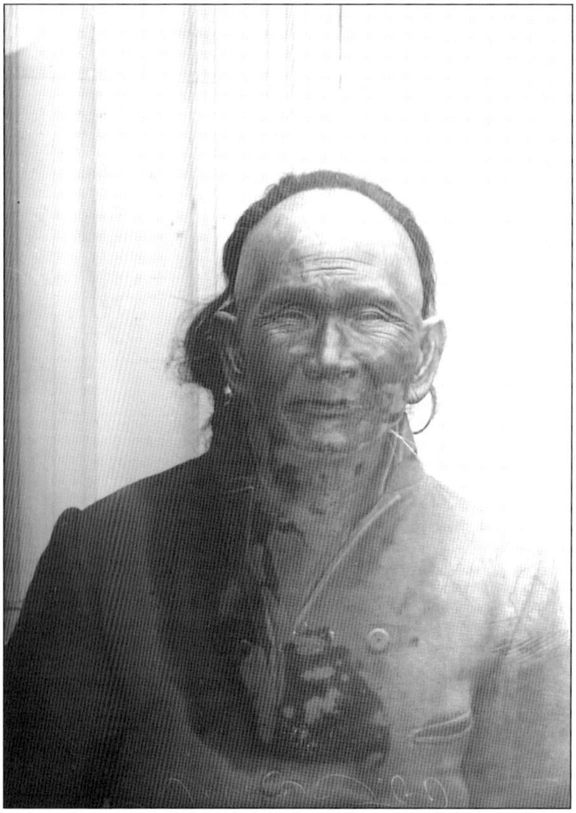

This photograph from 1900 is of an older Chinese man. The name "Gin Foo" is scratched in the emulsion of the photograph, and the picture shows the man wearing a contemporary wool jacket with a standup collar. A white paneled door can be seen in the background. An assortment of photographs of the residents of the Chinese Fishing Village can be found in the museum collection. (Courtesy Pacific Grove Museum of Natural History.)

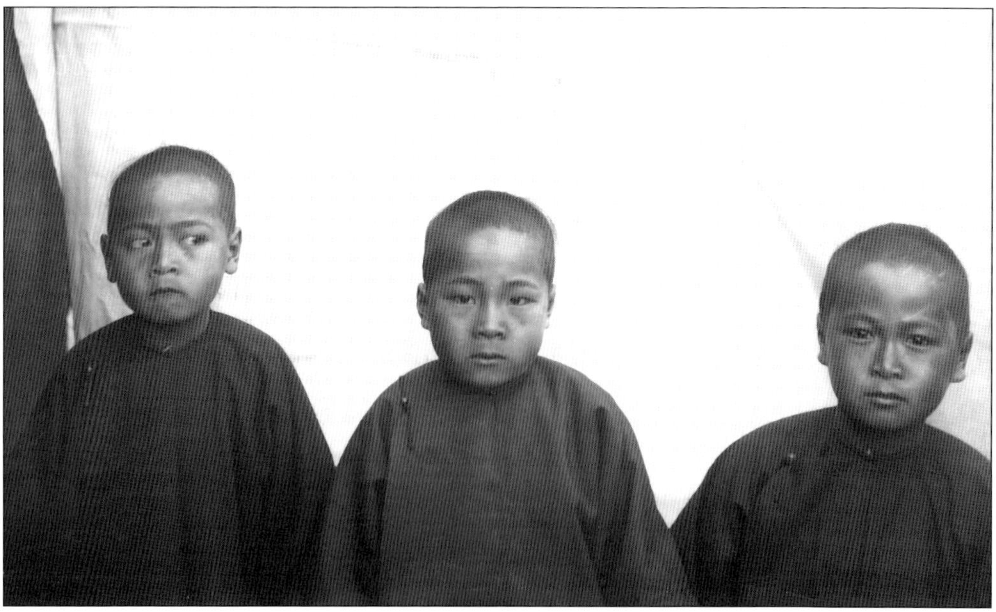

This photograph, also taken in 1900, offers a head-and-shoulders view of three Chinese children who lived in Pacific Grove's Chinese Fishing Village. A white cloth backdrop is visible in this image. The names "Chong Pat," "Chung Tai," and "Lui [or Loi]" Tai are scratched in the photograph's emulsion. (Courtesy Pacific Grove Museum of Natural History.)

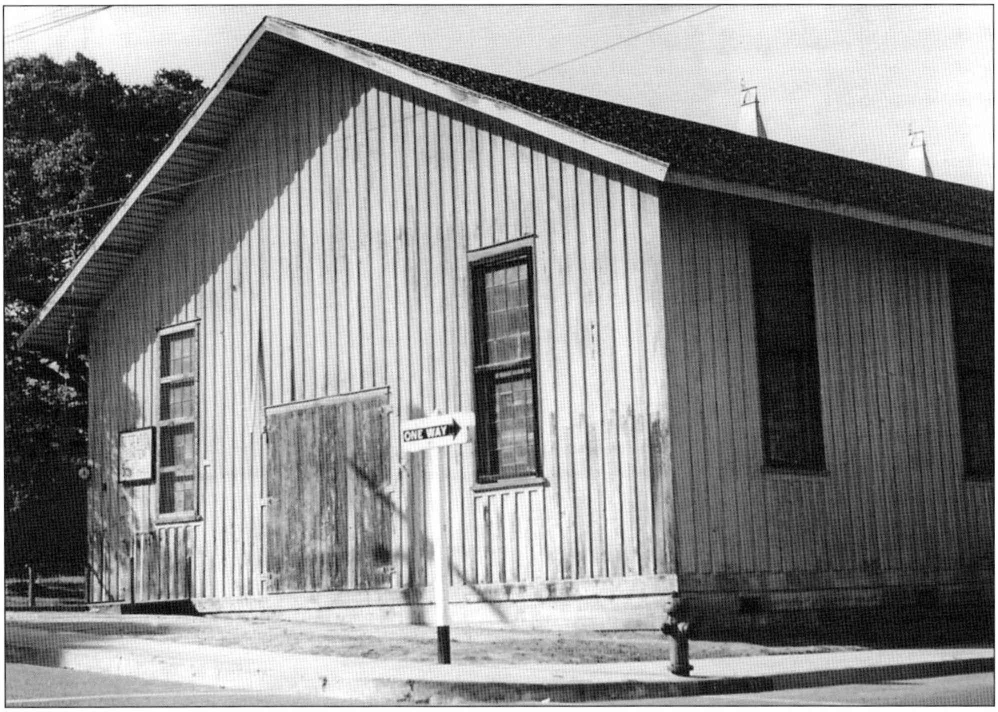

Chautauqua Hall is pictured prior to renovation work in 1988. The building, originally constructed in 1881 by the Pacific Grove Retreat Association, was declared a State Historical Landmark in 2005. Prior to the Pacific Grove Museum's opening in 1883, specimens for study were kept here. Today, the hall hosts regular special events. (Courtesy Pacific Grove Heritage Society.)

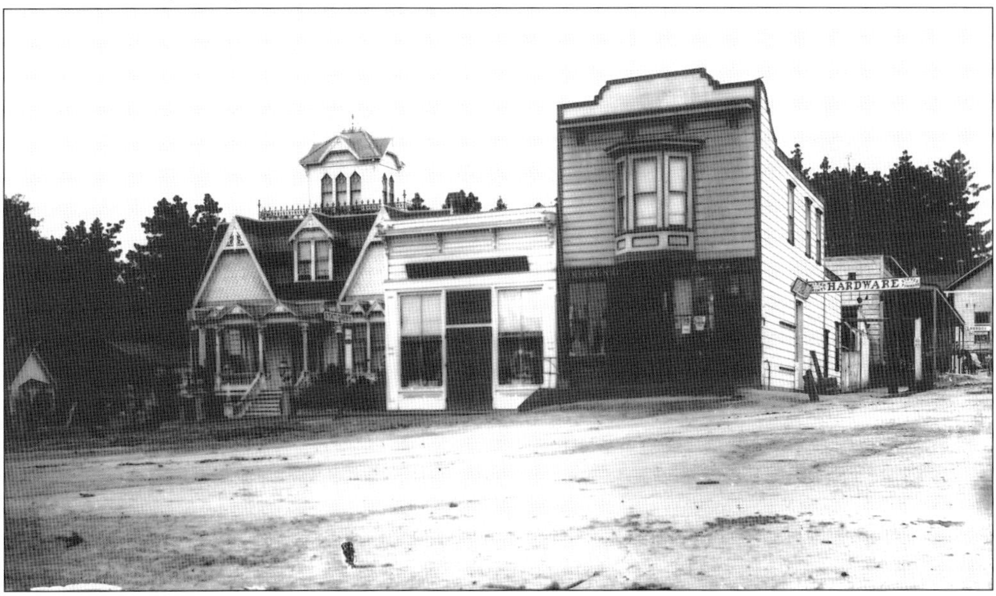

This C.K. Tuttle photograph from 1890, taken in the same decade the museum first opened, shows the business block of Lighthouse Avenue prior to paving. A cigar shop owned by T.K. Ray, a hardware store, and a laundry can also be seen in the image. Also visible is a fenced park and a rather ornate-looking home with a tower. (Courtesy Pacific Grove Museum of Natural History.)

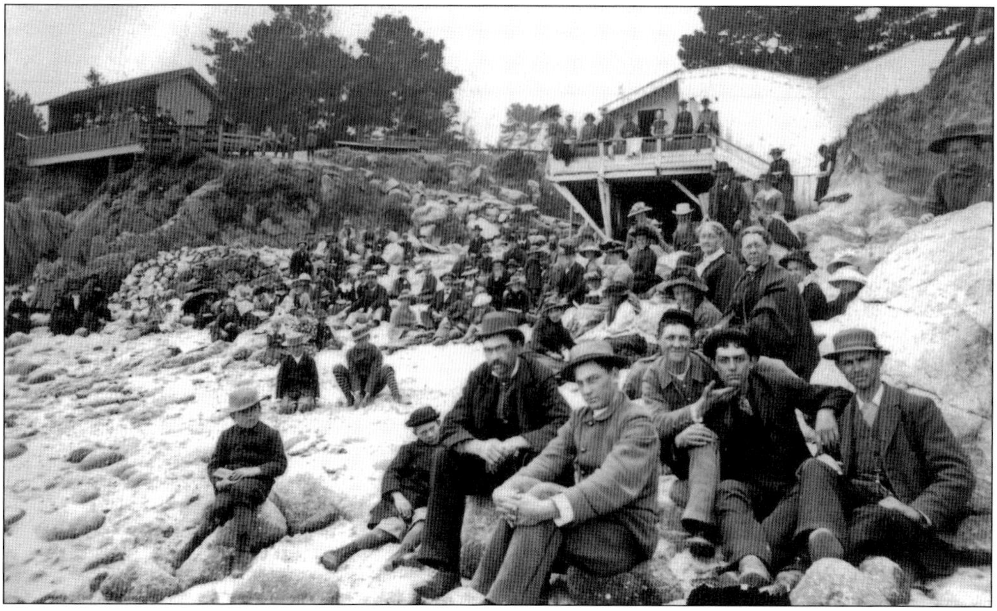

This 1885 photograph shows two bathhouses at Lovers Point in Pacific Grove with people in street clothes enjoying a day at the beach, including a group of young men hamming it up for the camera. Lovers Point remains a central location in the city for both residents and visitors. (Courtesy Pacific Grove Museum of Natural History.)

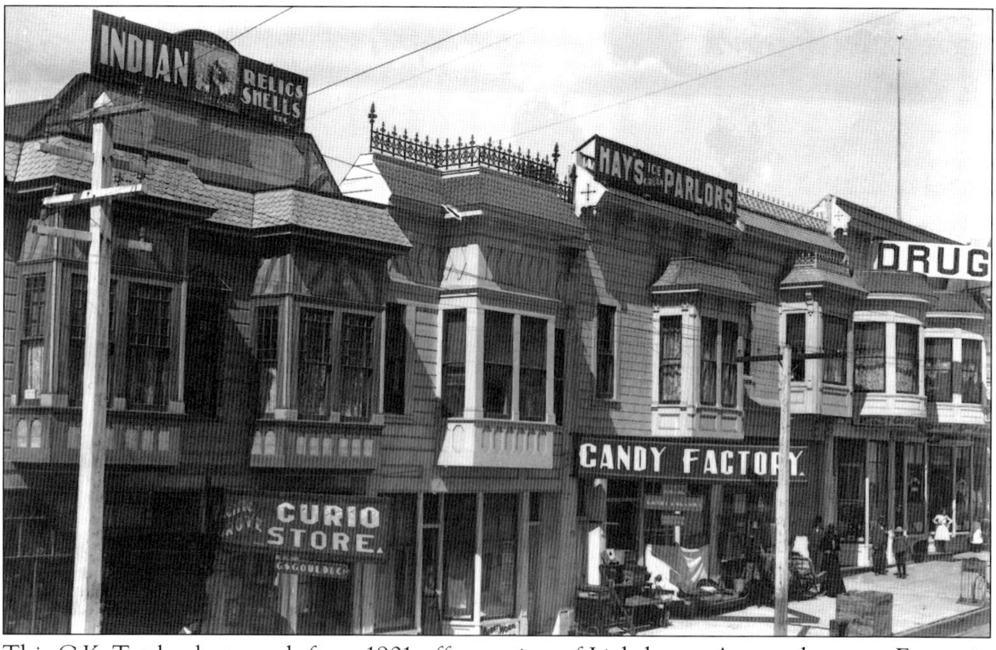

This C.K. Tuttle photograph from 1901 offers a view of Lighthouse Avenue between Fountain and Grand Avenues. From left to right are a curio store owned by G.S. Gould and Company, Burnt Works, Hayes Ice Cream Parlors, Candy Factory, a furniture store, and the Tuttle Pharmacy. Tuttle's own family members can also be seen in the photograph. (Courtesy Pacific Grove Museum of Natural History.)

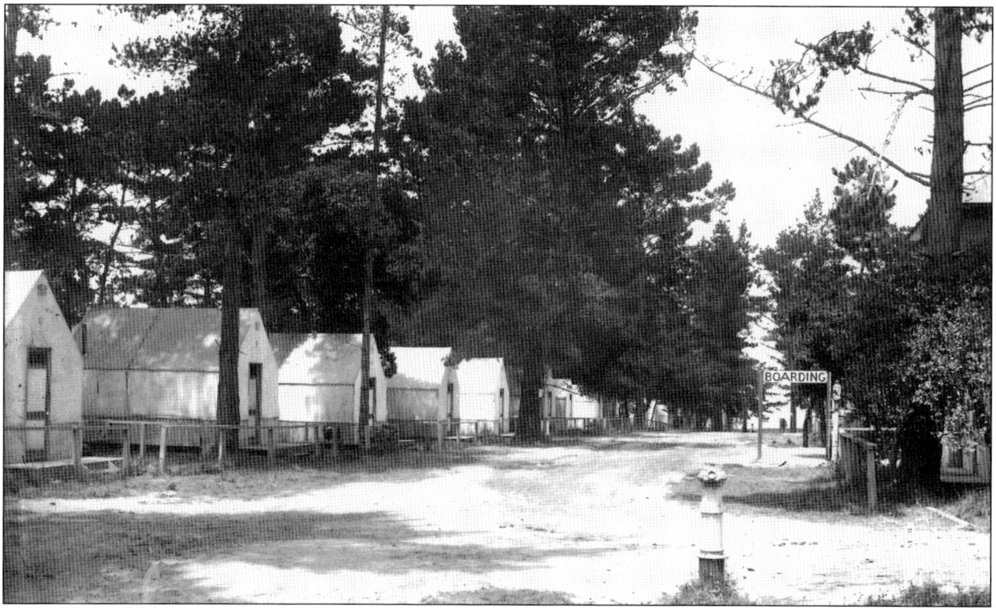

This picture from 1885, taken two years after the museum opened, shows several handsome tents on Forest Avenue in the retreat section of Pacific Grove. Also visible are an old streetlight, a fire hydrant, as well as a sign offering boarding. (Courtesy Pacific Grove Museum of Natural History.)

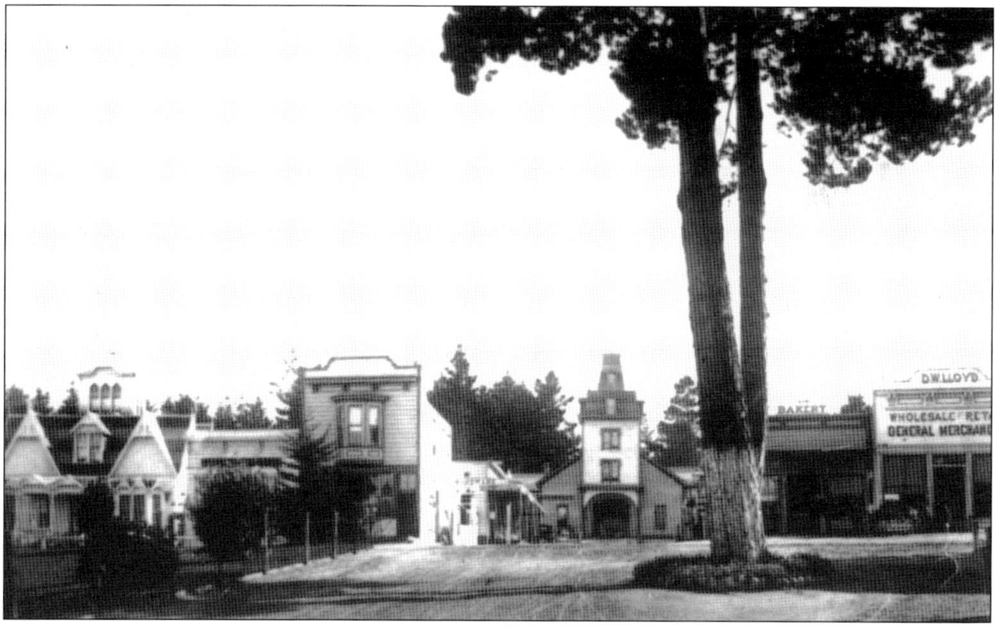

This 1890 photograph was taken at the corner of Lighthouse and Grand Avenues, looking south toward Mammoth Stables, located further up Grand Avenue. The museum could be found at the opposite end toward the beach at the time of this photograph. The image was captured on the grounds of the El Carmelo Hotel. A few shops are visible, including the Tuttle Pharmacy and a bakery. (Courtesy Pacific Grove Museum of Natural History.)

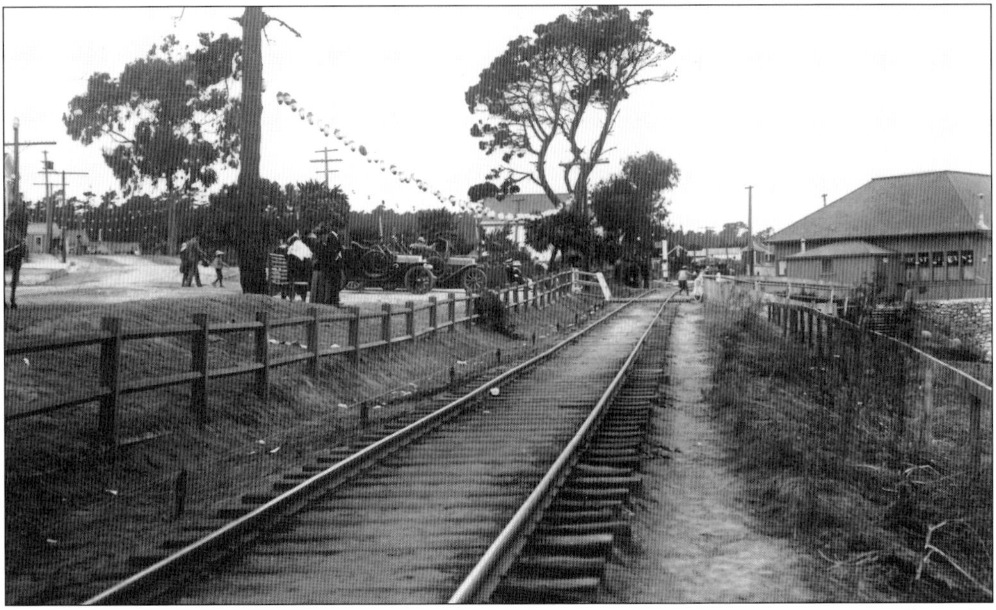

Visible in this image from 1915 are the Southern Pacific train tracks near Lovers Point along the Pacific Grove shoreline. Also visible in this image is a string of lanterns hung in honor of the city's annual Feast of Lanterns, an event that is still held at Lovers Point today. (Courtesy Pacific Grove Museum of Natural History.)

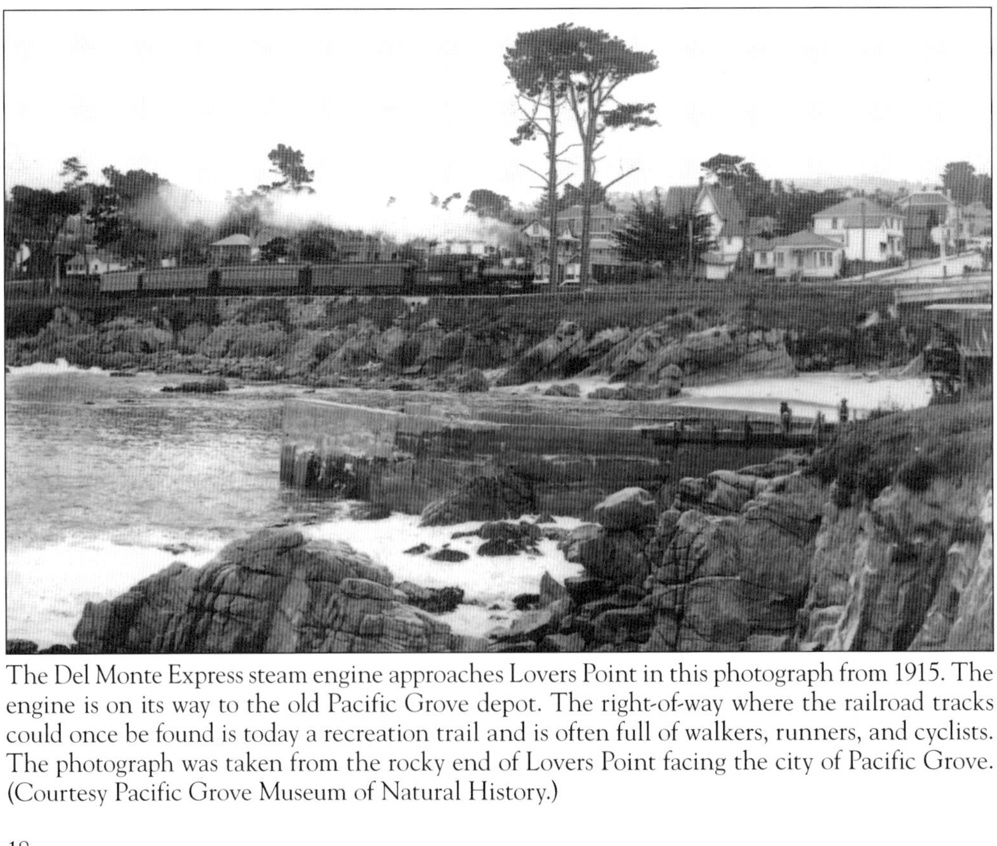

The Del Monte Express steam engine approaches Lovers Point in this photograph from 1915. The engine is on its way to the old Pacific Grove depot. The right-of-way where the railroad tracks could once be found is today a recreation trail and is often full of walkers, runners, and cyclists. The photograph was taken from the rocky end of Lovers Point facing the city of Pacific Grove. (Courtesy Pacific Grove Museum of Natural History.)

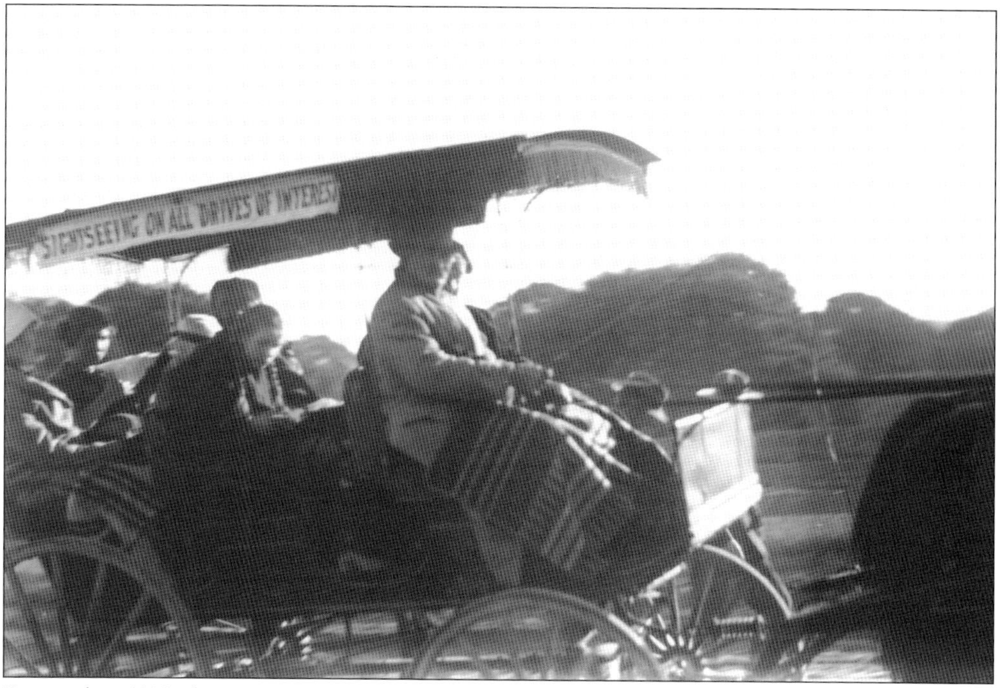

Pictured in 1910, this tourist carriage full of happy passengers likely stopped at the museum as it carted tourists through the Pacific Grove area. A sign on the top of the carriage reads, "Sightseeing on all drives of interest." Today, buses often make their way to the museum for special tours. (Courtesy Pacific Grove Museum of Natural History.)

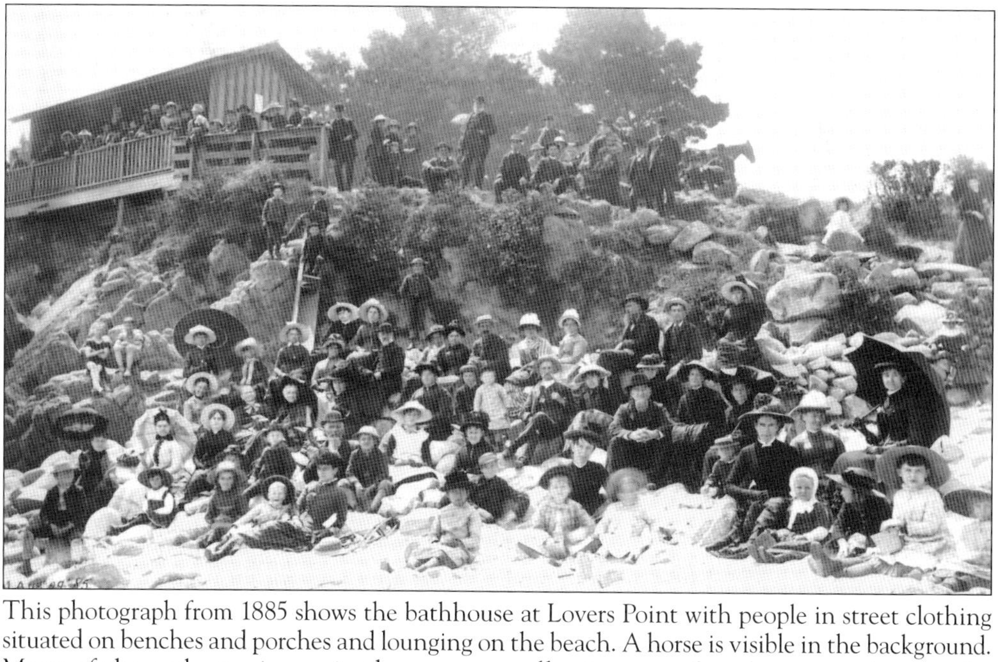

This photograph from 1885 shows the bathhouse at Lovers Point with people in street clothing situated on benches and porches and lounging on the beach. A horse is visible in the background. Many of the early specimens in the museum collection were found on excursions to the shoreline along Pacific Grove, including plants and shells. (Courtesy Pacific Grove Museum of Natural History.)

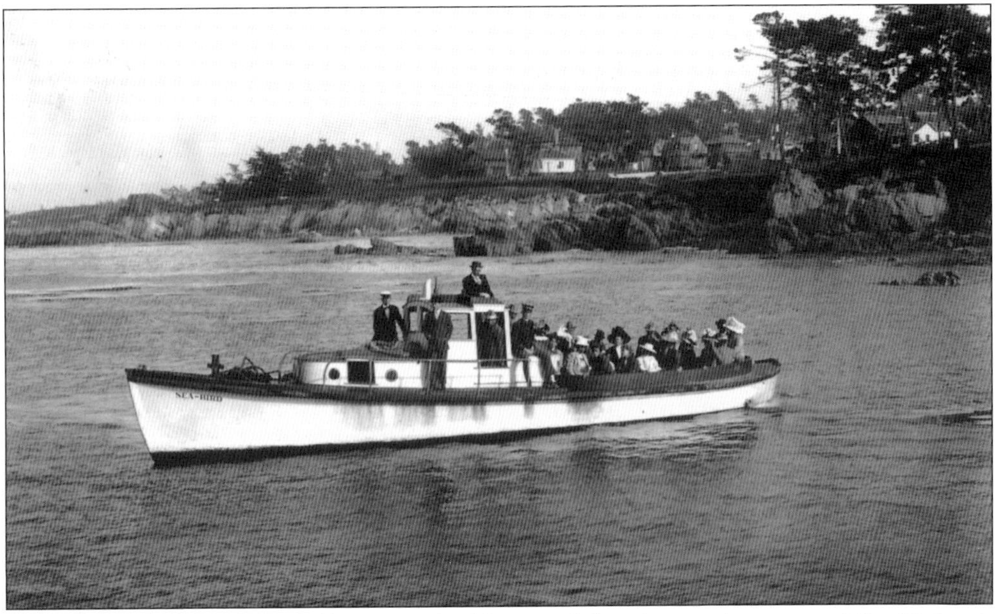

This photograph from 1910 offers a wonderful view of the pleasure boat named *Sea-Bird* as it cruises over Monterey Bay with a load of passengers. The vessel is pictured near Ocean View Boulevard in Pacific Grove. Many homes are visible in the background. Visitors and residents alike often flock to the area for its beautiful natural views. (Courtesy Pacific Grove Museum of Natural History.)

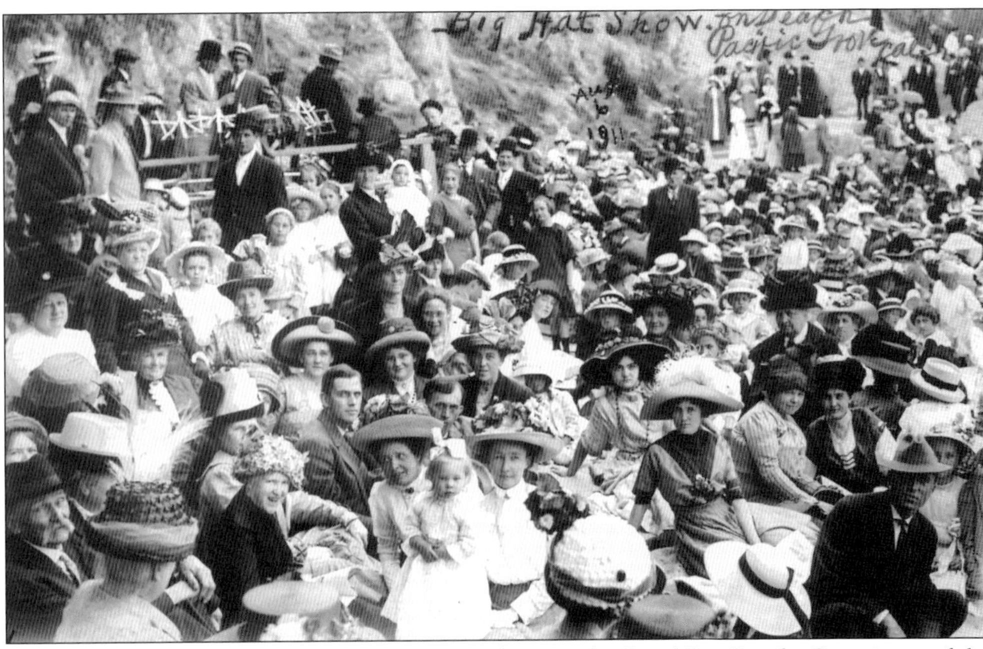

This image from 1911 celebrates the "Big Hat Show on the Beach" in Pacific Grove, one of the many festive events residents would hold in the small seaside community. This event featured a band concert and several people can be seen in street clothing flocking to the event. (Courtesy Pacific Grove Museum of Natural History.)

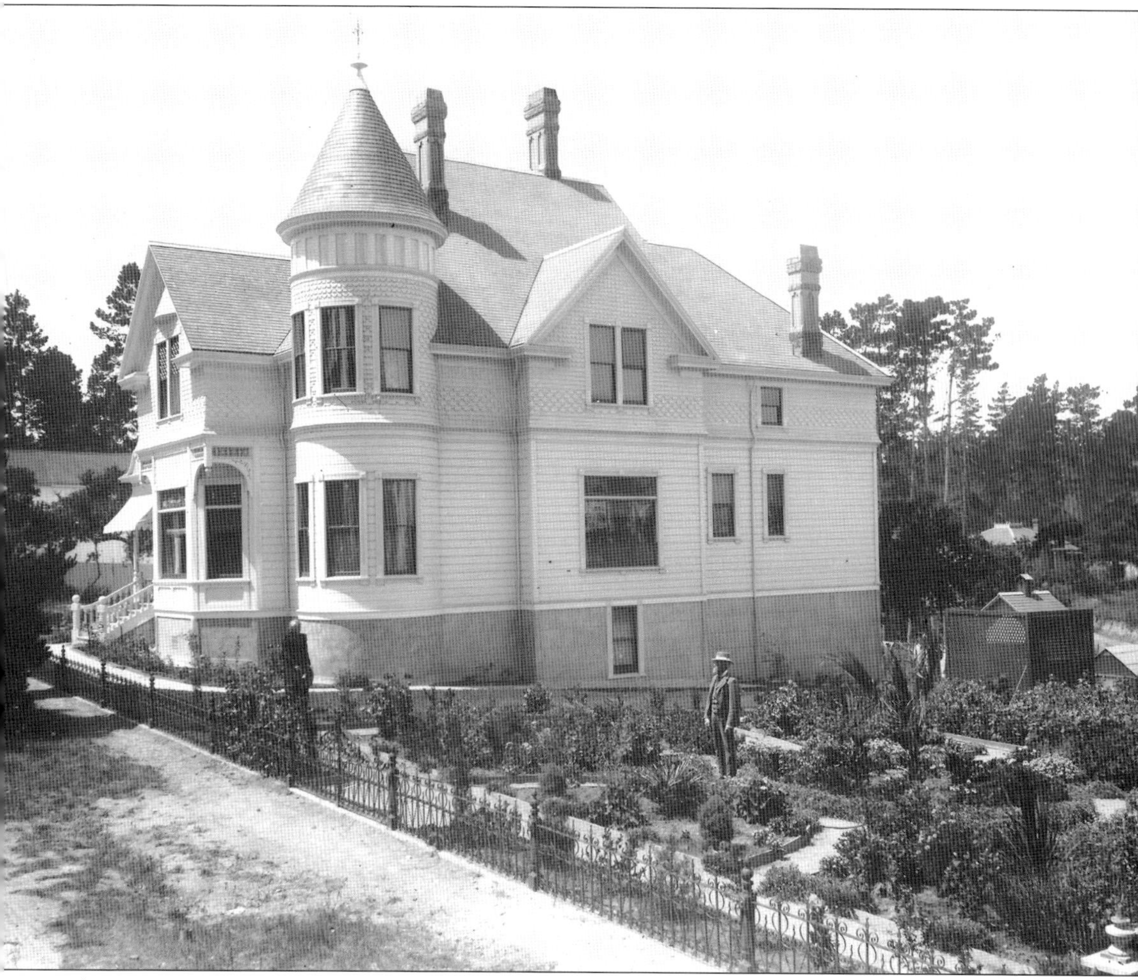

The austere yet ornate Trimmer Hill is visible in this 1896 photograph. The beautiful Pacific Grove home was located at Sixth Street and Laurel Avenue. Dr. O.S. Trimmer, with a full beard and under a hat, is looking toward his wife, Rhoda. Also visible is what appears to be a large, plentiful garden. The Trimmers' young niece Emily married local photographer and pharmacist C.K. Tuttle. Tuttle photographed many of the images of old Pacific Grove throughout the years, and many of which are part of the museum collection. (Courtesy Pacific Grove Museum of Natural History.)

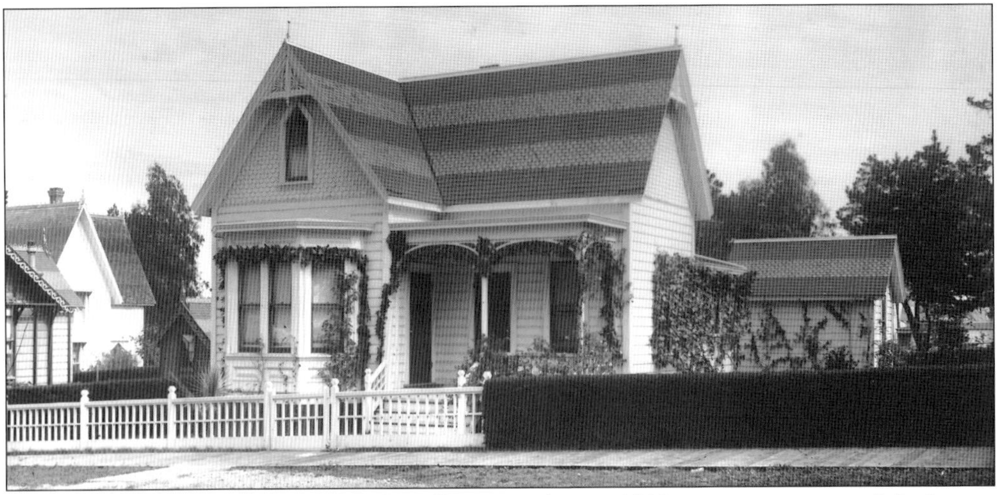

This photograph from the C.K. Tuttle collection, taken in 1900, is representative of the homes found in the neighborhood surrounding the museum. This particular home is located at 119 Forest Avenue and was known as the Wilbur House. Built in 1885, the home features vines looping around the porch arches, a Gothic front window, fish-scale siding, a bay window, and a wood plank sidewalk outside of the two-story home. (Courtesy Pacific Grove Museum of Natural History.)

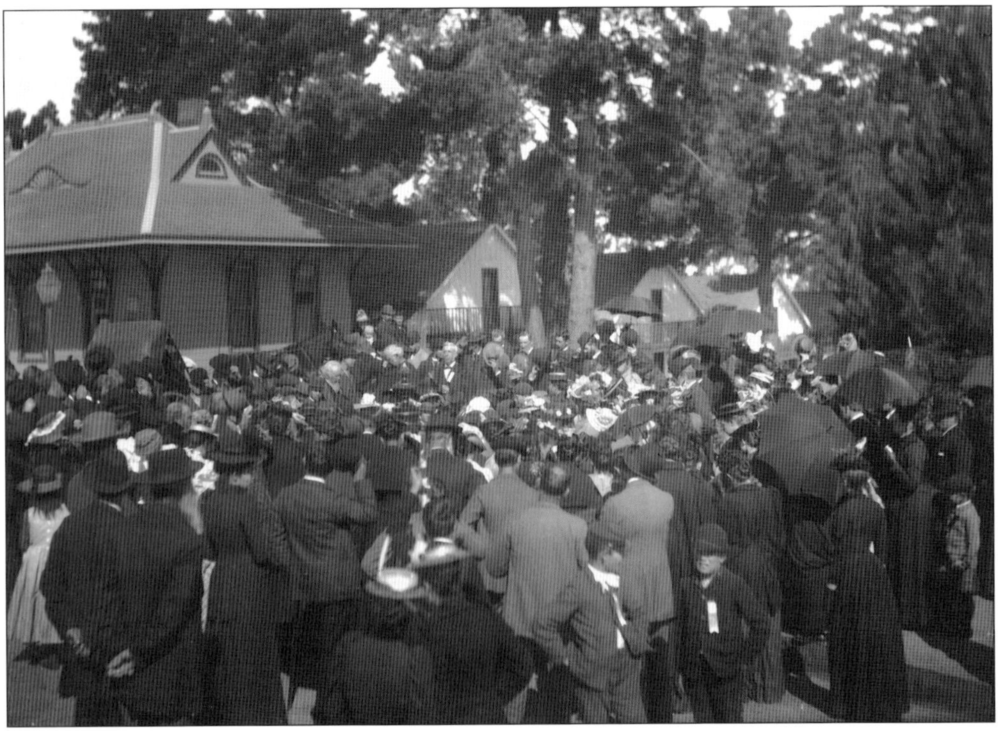

In this image from the C.K. Tuttle collection, a crowd has gathered on Lighthouse Avenue in May 1901, during the visit of Pres. William McKinley to Pacific Grove. McKinley is visible near the water fountain with his head bowed. The Pacific Improvement Company office building is also seen in the image. McKinley visited the area during the 34th Grand Army of the Republic encampment. He would be shot later that year in New York and soon succumbed to his injuries. (Courtesy Pacific Grove Museum of Natural History.)

Two

The Early Museum

In 1881, two years before the creation of the Pacific Grove Museum, crews with the Pacific Improvement Company constructed at the corner of Sixteenth Street and Central Avenue, then known as Grove Street, what is today referred to as Chautauqua Hall. The hall is where members of the Chautauqua Literary and Scientific Circle Assembly taught classes, stored their summer tents, and held meetings. The hall also stored natural history specimens, which included an herbarium of locally collected flora. This desire for educational natural history objects directed efforts to establish a museum in Pacific Grove, though the building itself would not be ready until two years later in 1883. A petition signed by notable Chautauqua Assembly members, including Lucy M. Washburn, Mary Norton, Josiah Keep, and Prof. H.B. Norton, and sent to then PIC secretary Frank Douty led to the donation of a small building at the location now home to Jewell Park. Kate Coffin also helped influence the PIC in their donation. Prior to opening, however, members sought collection items in the form of dried plants, minerals, geological specimens, and illustrations "in all departments of zoology." Requests for collection items continued years after opening the museum as well, as evidenced in an article in the *Pacific Rural Press* in 1886. During this time, Pacific Grove was steadily growing larger. Artists were flocking to experience the beauty of the area, as were famous writers such as Robert Louis Stevenson and, later, John Steinbeck. Visitors to the area also noticed the large number of monarch butterflies that would overwinter in the area, which added to Pacific Grove's gentle charm. In 1939, city officials made it a misdemeanor to molest a butterfly. By 1900, the museum owned an impressive collection of approximately 2,100 natural history objects. It was not until 13 years later, however, that talk began to surface of transferring ownership of the museum to the City of Pacific Grove, which incorporated 24 years before in 1889. Talk continued into 1916, and an election decided the matter once and for all when the museum became a part of the City of Pacific Grove in 1917.

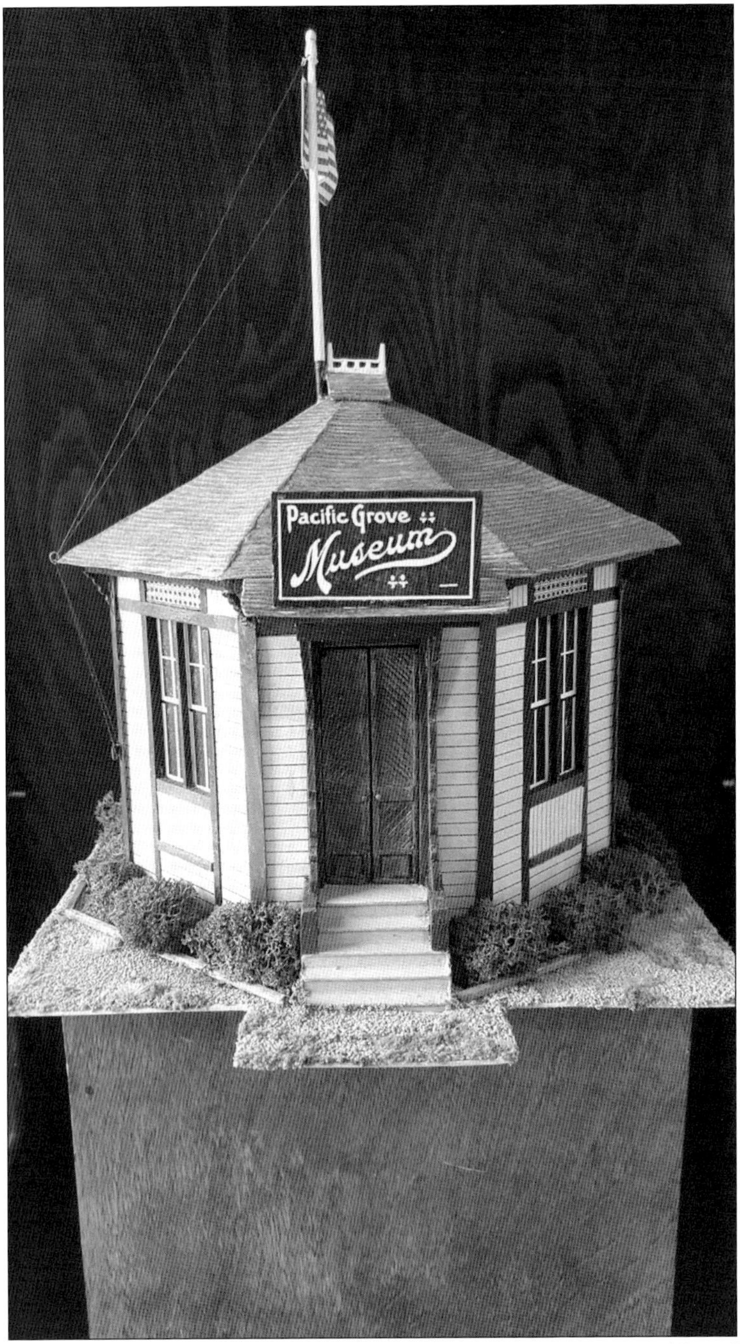

This model replica of the original Chautauqua octagon museum was created by talented model artist Snick Farkas. The museum keeps several model replicas in its collection. These include the replica of a pueblo structure, a model replica of the modern Pacific Grove Museum in its current form, and two models of the Chinese Fishing Village, which are currently on display. The octagon museum model was made to illustrate the size and shape of the early building to give museum guests an idea of the small accommodations available to specimens in 1883. (Courtesy Pacific Grove Museum of Natural History.)

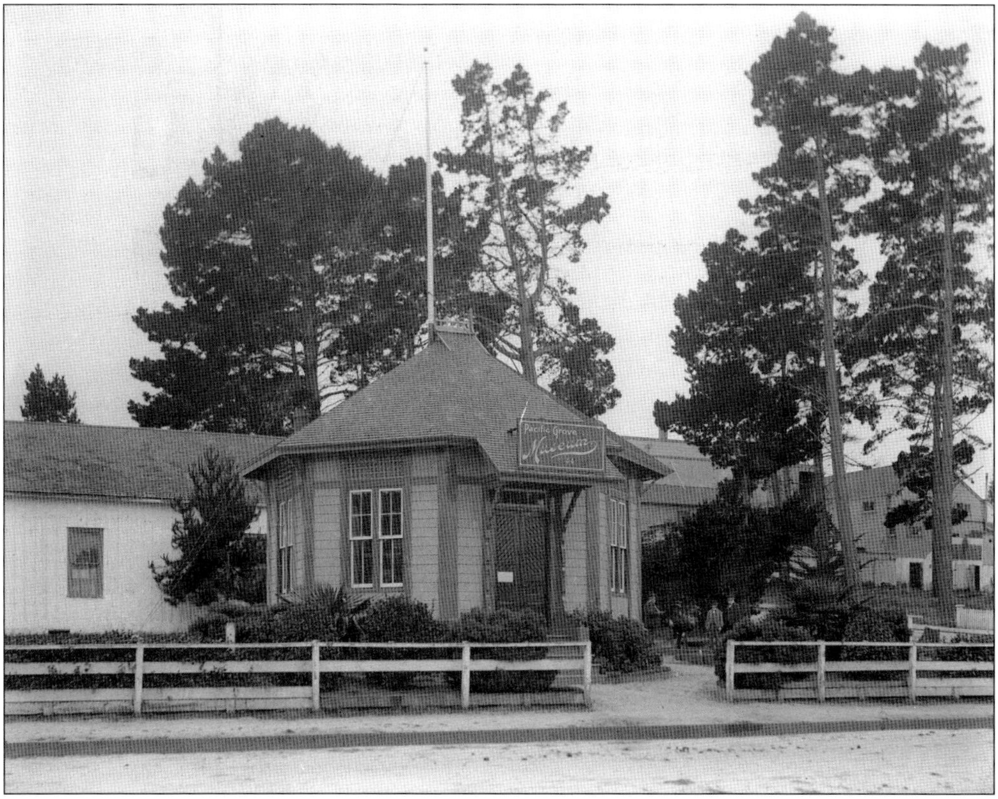

This c. 1900 image from the C.K. Tuttle collection shows the original octagon-shaped Pacific Grove Museum. The view, seen from Central Avenue, formerly known as Grove Street, offers a glimpse of Forest Avenue. Men are also visible in the photograph as they play croquet nearby. The Chautauqua Literary and Scientific Circle used the structure during their two-week summer retreats. It was replaced by the current museum building in 1932 and has grown over the years. (Courtesy Pacific Grove Museum of Natural History.)

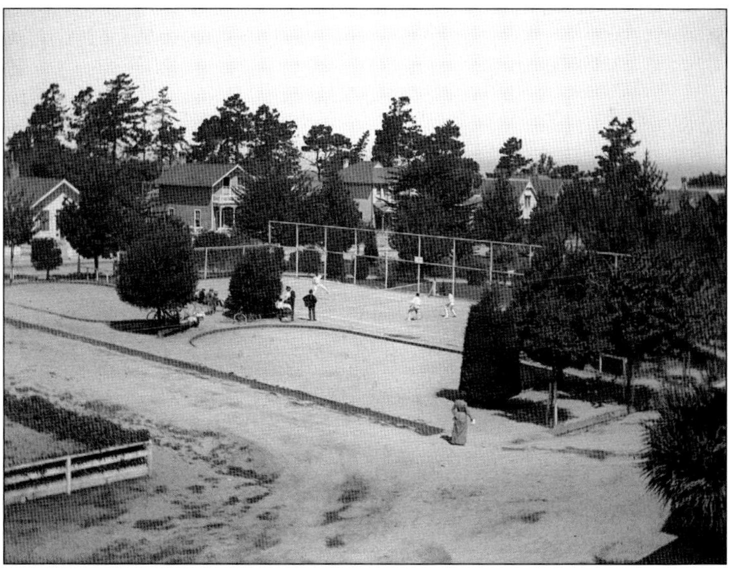

In this August 20, 1900, photograph from the C.K. Tuttle collection, a doubles match is in progress at the Jewell Park tennis court. The image was captured from the roof of the El Carmelo Hotel, looking northwest. A portion of the museum grounds is also visible in the image. (Courtesy Pacific Grove Museum of Natural History.)

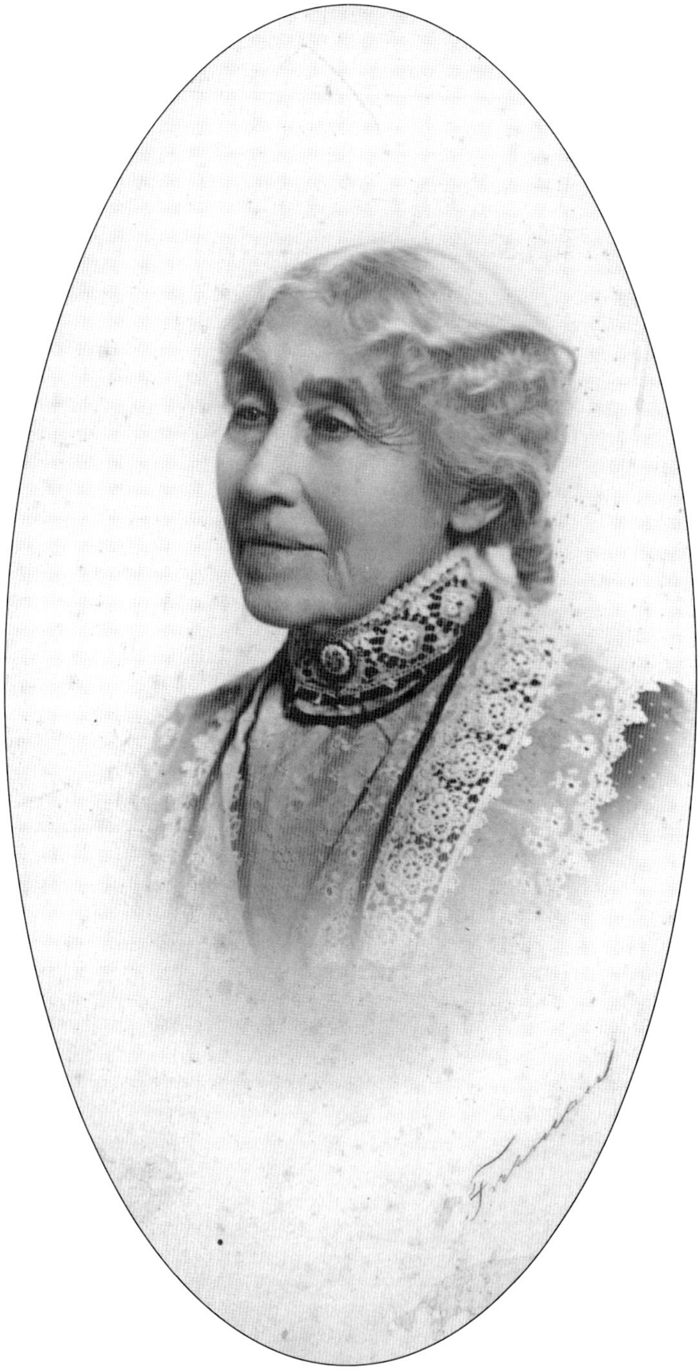

This is a 1915 portrait of the museum's first curator, Mary E.B. Norton. Born on November 30, 1832, in New York, Norton was educated at the Rockford Seminary in Illinois and later in Berlin, Germany. She taught in the Midwest for 18 years, before joining her brother, Prof. Henry Brace Norton, at the State Normal School at San Jose in 1878, a school now known as San Jose State University. In the late 1800s, her students presented her with a microscope, which can now be found in the museum's collection. (Courtesy Pacific Grove Museum of Natural History.)

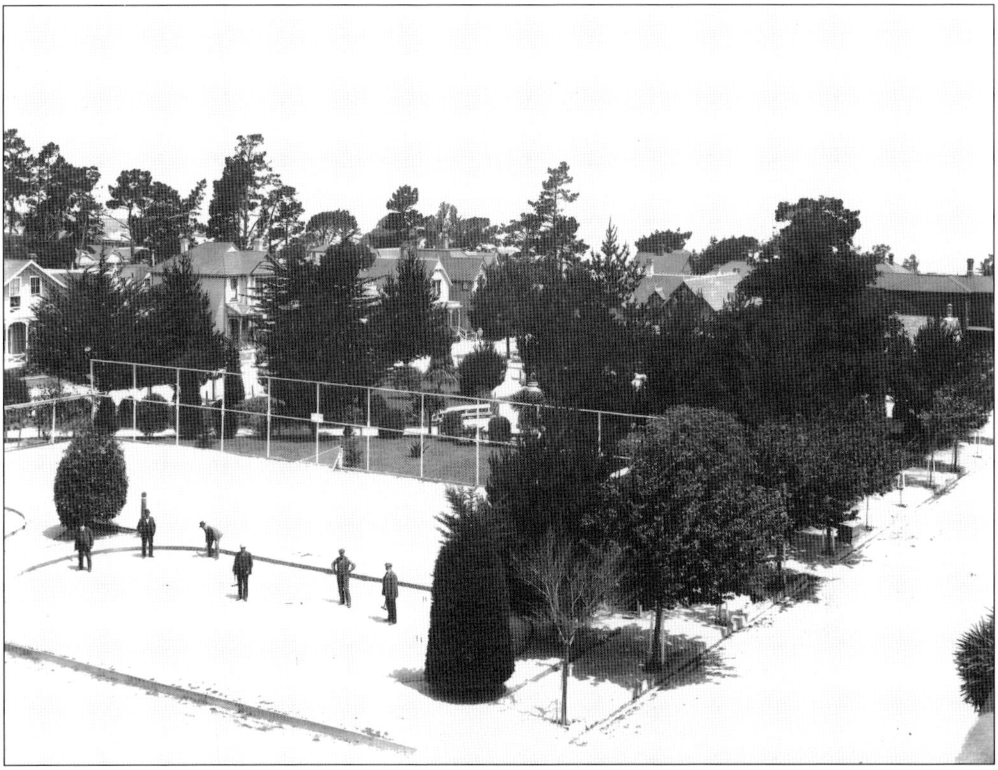

This 1905 photograph offers a view of what is today called Jewell Park. A roque game is in progress. Roque was a game like croquet but played on concrete with smaller mallets. Glimpses of Jewell Cottage can be seen through the trees. This photograph was taken from the roof of the nearby El Carmelo Hotel. (Courtesy Pacific Grove Museum of Natural History.)

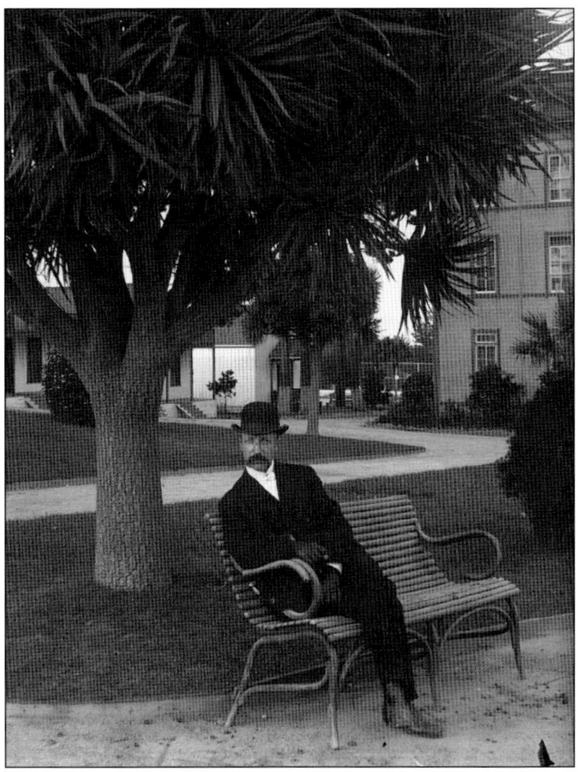

This image from the C.K. Tuttle collection shows a derby-hatted man sitting on a park bench in the lush garden in front of Pacific Grove's El Carmelo Hotel, with the hotel's cabins visible at the left. The cabins housed the Pacific Grove Women's Civic Improvement Club, which was founded in 1903. The cabins were later used as an early add-on to the Pacific Grove Museum of Natural History's original building. (Courtesy Pacific Grove Museum of Natural History.)

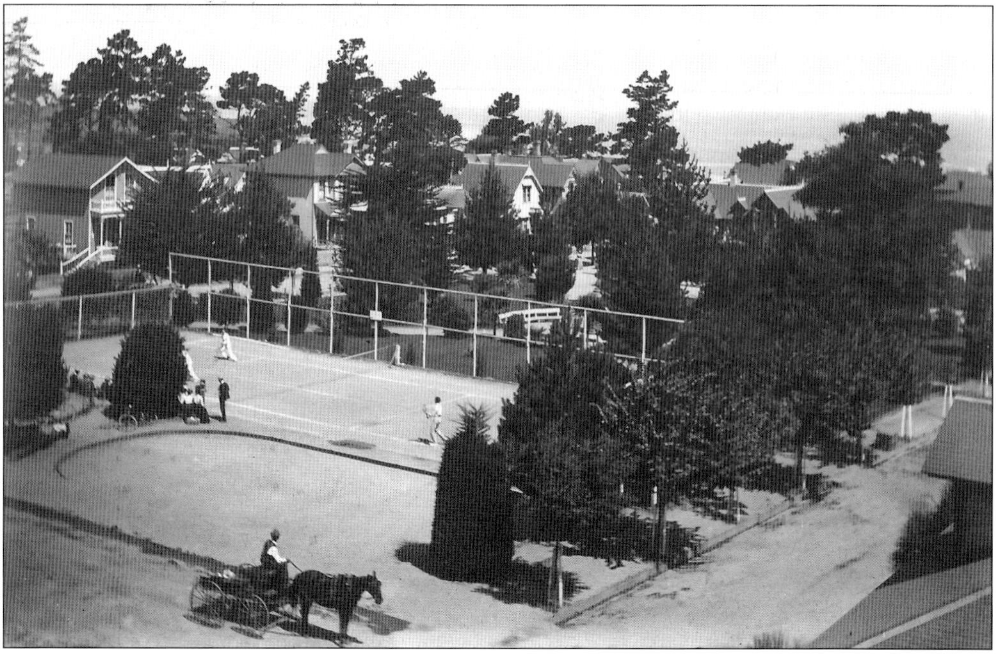

This image from 1900 shows what is now Jewell Park, located at the corner of Forest and Central Avenue and the location of the museum. A game is in progress on the tennis court in the background. Like other images of Jewell Park, this one was taken from the second floor of the El Carmel Hotel. A horse and wagon is also visible at the corner of Grand Avenue. (Courtesy Pacific Grove Museum of Natural History.)

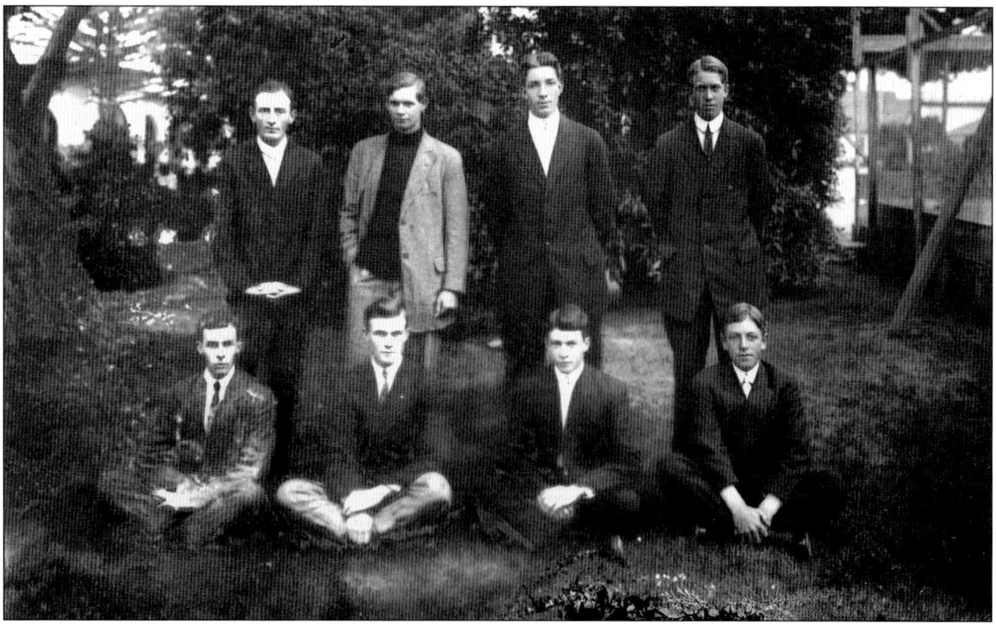

This 1910 photograph shows eight young men posing together in what is today called Jewell Park. The library building is visible in the background at the left of the image. Fencing around the tennis court, near the museum's old location, is also visible. The eight young men may have been posing for a graduation photograph. (Courtesy Pacific Grove Museum of Natural History.)

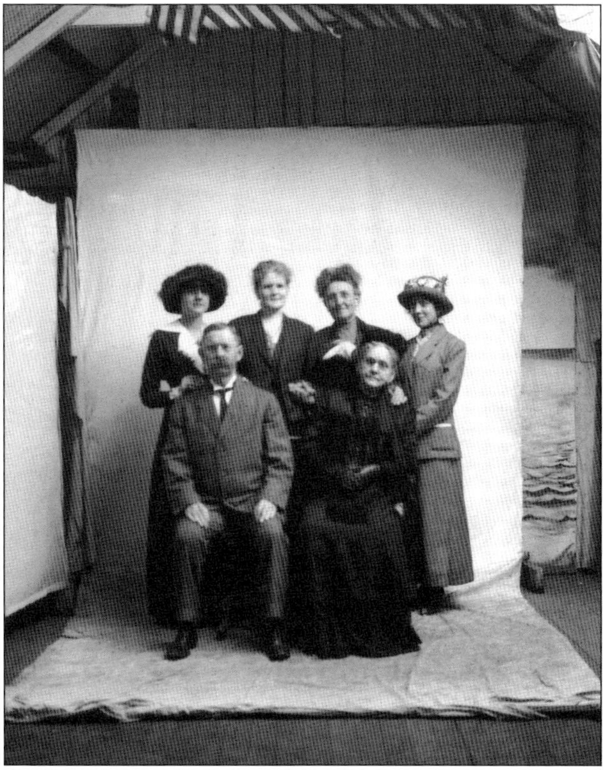

This 1910 studio portrait shows five women and one man, with the interior of the photography studio visible. The woman seated at the front and right resembled Mary Norton, the museum's first curator. A painted backdrop of Lovers Point is visible behind the group. (Courtesy Pacific Grove Museum of Natural History.)

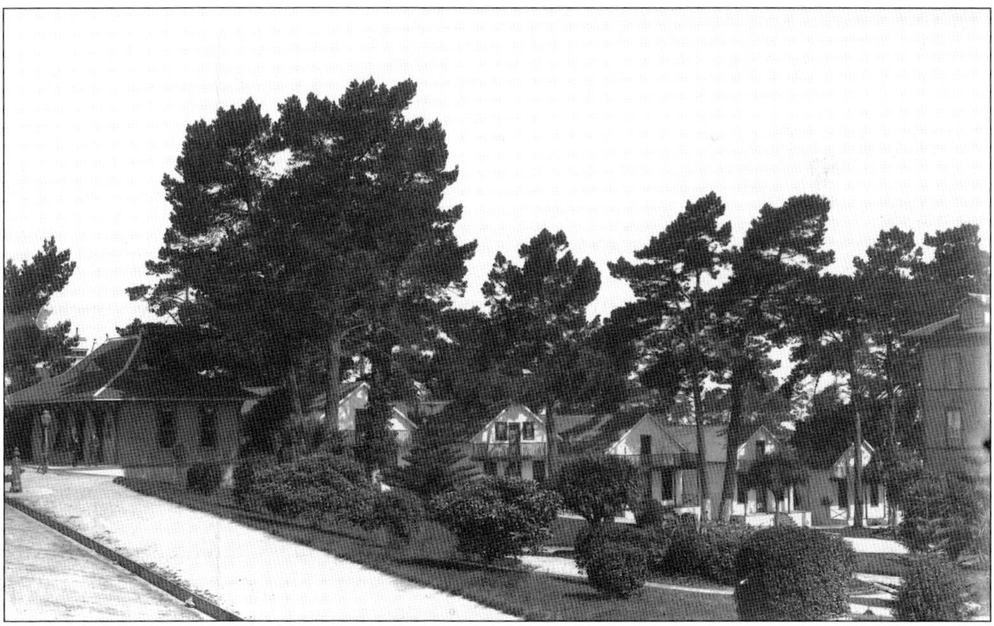

This 1910 photograph shows the Pacific Grove Hotel, earlier known as the El Carmelo Hotel, and its front garden, as seen from Lighthouse Avenue. The office of the Pacific Improvement Company, which helped facilitate the museum's first official building, can also be seen, complete with an "eyebrow" window, a design element popular at the time. Hotel cabins are also visible in the image. (Courtesy Pacific Grove Museum of Natural History.)

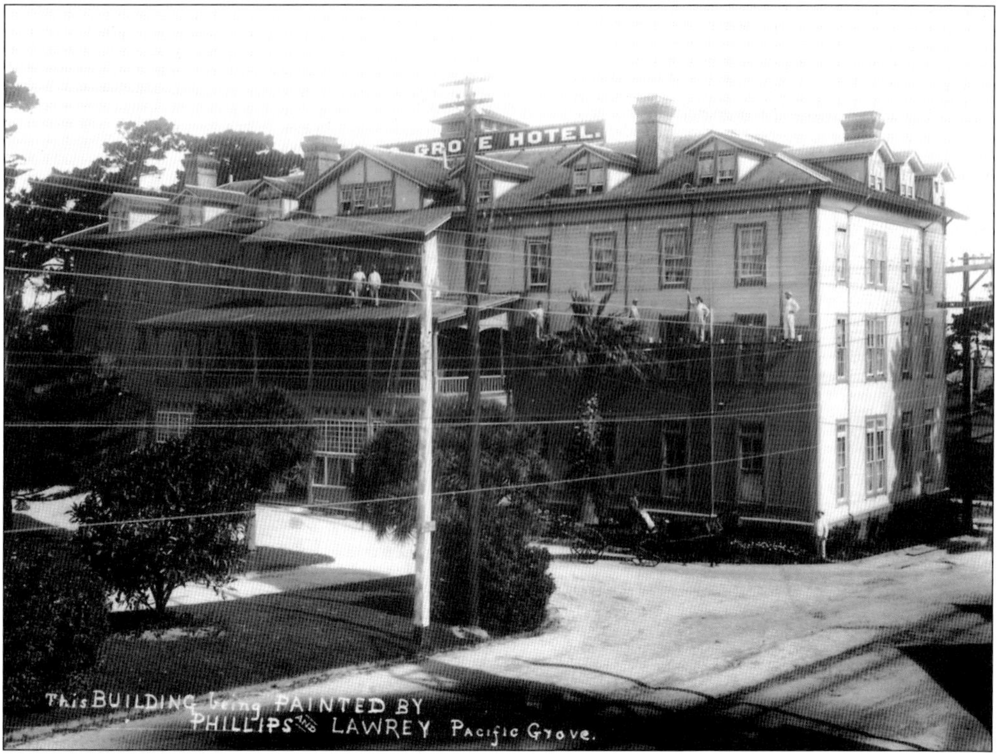

Pictured here in this 1910 image is the Pacific Grove Hotel as it is undergoing painting. The hotel, found directly across the park from the museum, was originally called the El Carmelo Hotel. Visible in this image are rails for the streetcar and painting crews with Phillips and Lawrey. The site would later become the Holman Department Store. (Courtesy Pacific Grove Museum of Natural History.)

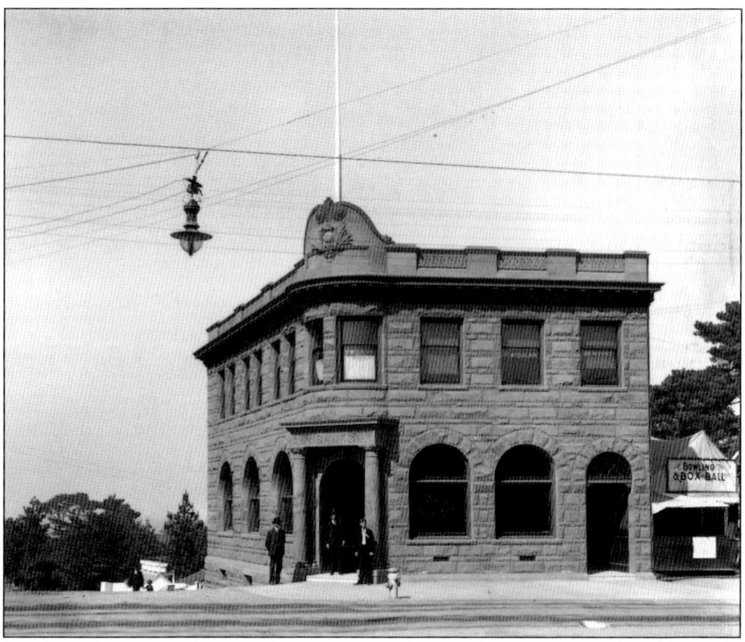

This C.K. Tuttle collection photograph, taken in 1905, offers a view of the Bank of Pacific Grove Building with two men and a boy standing in the front. Bank president and first mayor of Pacific Grove O.S. Trimmer is the man on the left of the photograph. The land behind the photograph would become home to the current museum. (Courtesy Pacific Grove Museum of Natural History.)

This photograph from 1882 shows a bearded Oliver Smith "O.S." Trimmer at the age of 48. Trimmer was born in 1883 and was a longtime resident of Pacific Grove, not to mention the city's first mayor. Trimmer also lived in Salinas, California, where he practiced medicine and served as the coroner for Monterey County. His family moved to a small cottage at the corner of Sixth Street and Laurel Avenue in 1888, five years after the museum first opened. (Courtesy Pacific Grove Museum of Natural History.)

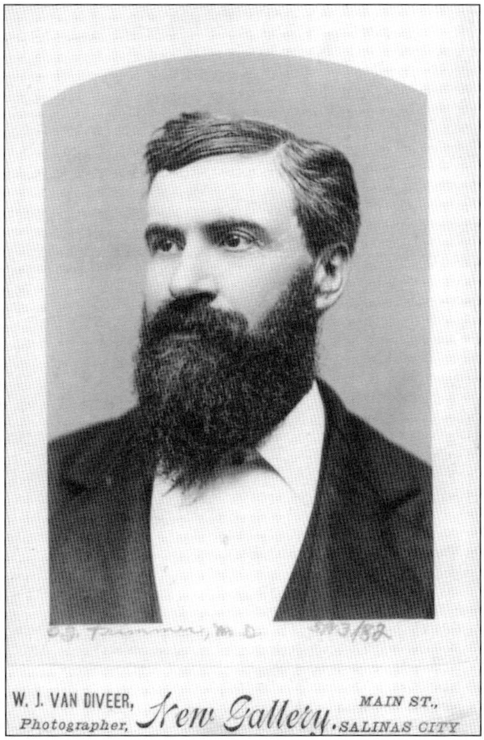

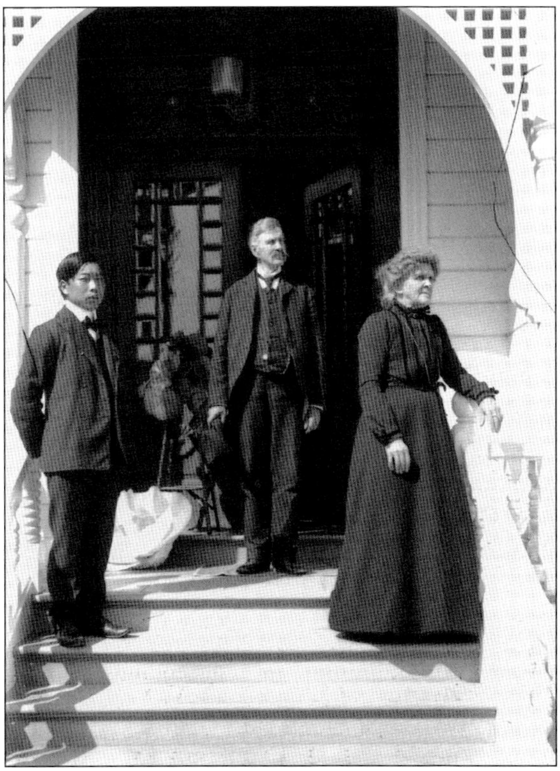

Trimmer was comfortably at home in Pacific Grove by the time this 1905 image was captured. Standing on the front steps of their Trimmer Hill home, from left to right are Kimura "Kimmie" Kangelo, Dr. Trimmer, and Rhoda Trimmer. Besides being the first mayor of Pacific Grove, Trimmer worked throughout the community in other areas, including as a board member for the museum. (Courtesy Pacific Grove Museum of Natural History.)

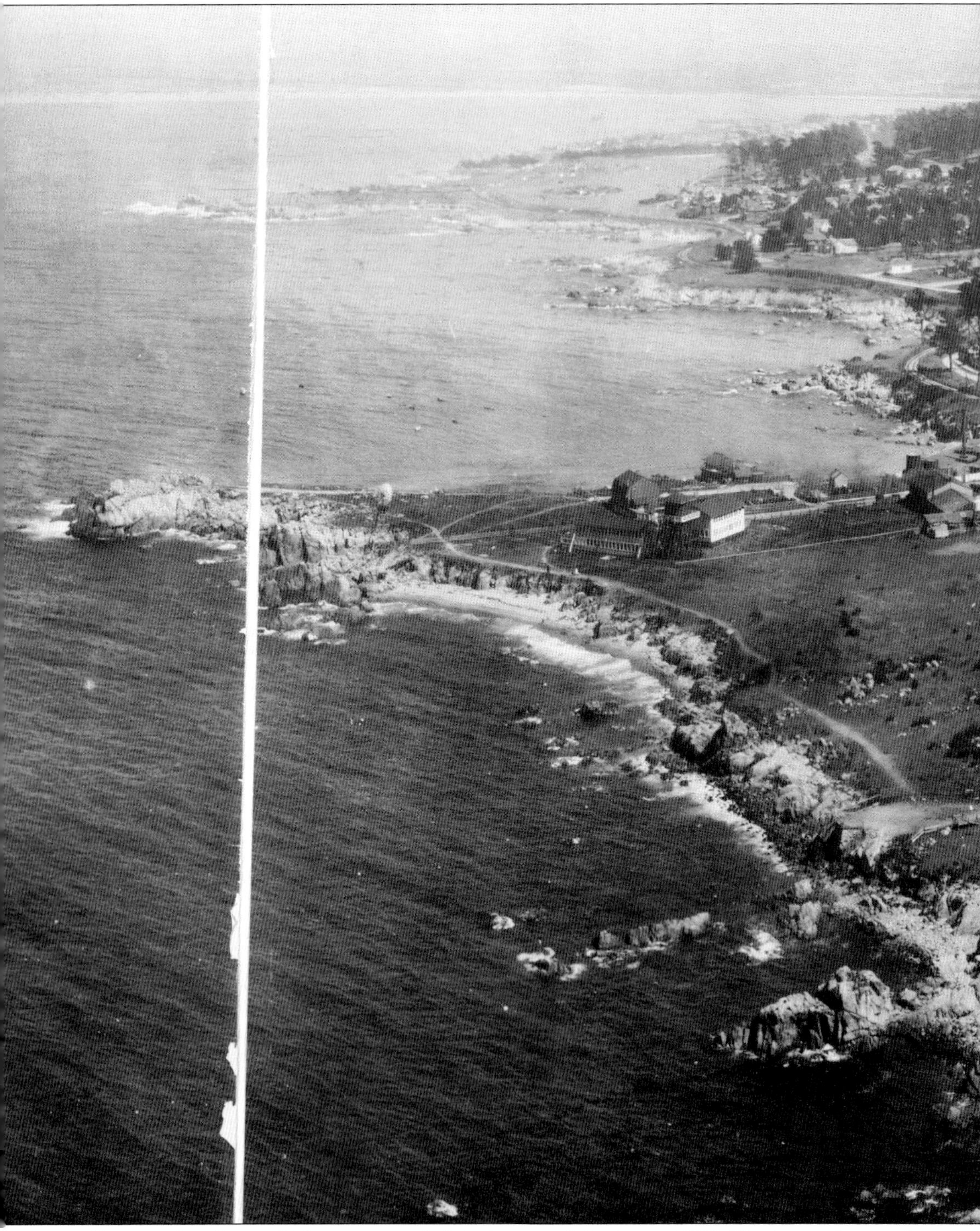

Photographer George R. Lawrence took this photograph in August 1906. The image shows a sky-high view of Pacific Grove. Visible are several buildings, including the museum. Lawrence was famous for flying his cameras on kites to capture great images. Lawrence used his kites to capture images of San Francisco following the devastation caused by the 1906 quake. One particular San

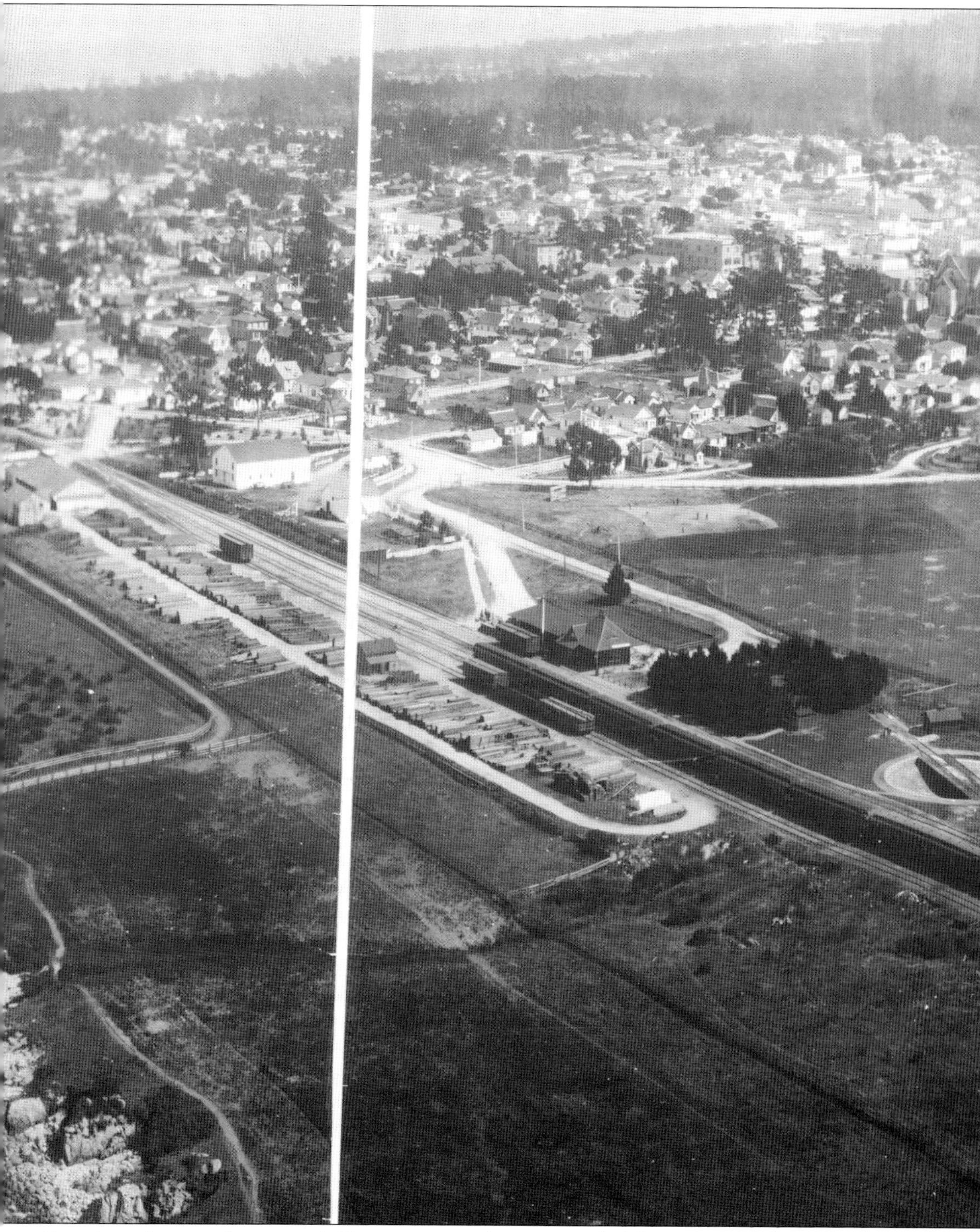

Francisco photograph was carried by newspapers from one end of the country to the other. The image reportedly earned Lawrence $15,000, quite a sum at that time. Besides Pacific Grove and San Francisco, he also used kites to carry his camera over other local communities, including Salinas and Santa Cruz. (Courtesy Library of Congress.)

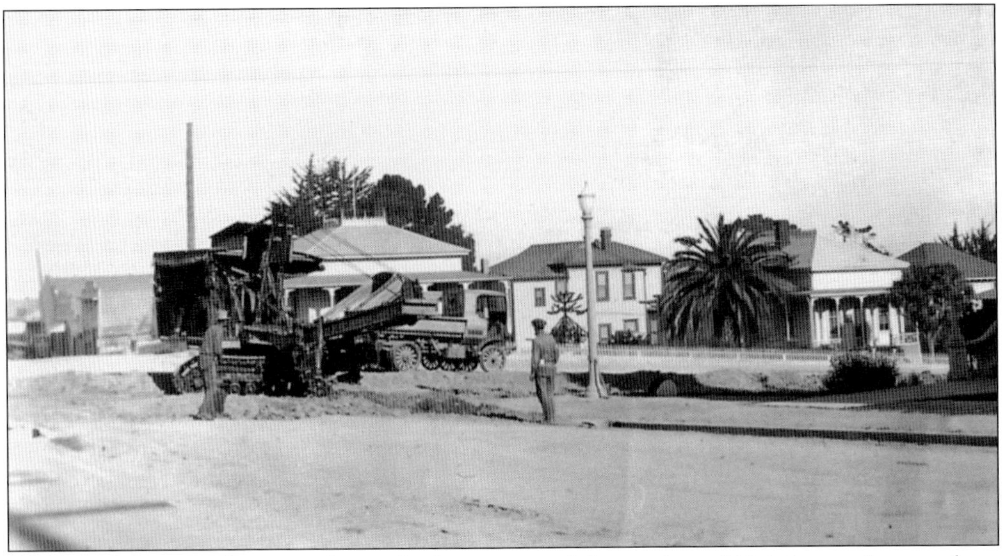

This 1924 C.K. Tuttle image shows the beginning of the paving of Lighthouse Avenue from Fountain Avenue looking north. Visible in the photograph is Ed Rickett's lab and the house that belonged to C.K. Tuttle's mother. An excavator and truck can also be seen in the corner of the image. (Courtesy Pacific Grove Museum of Natural History.)

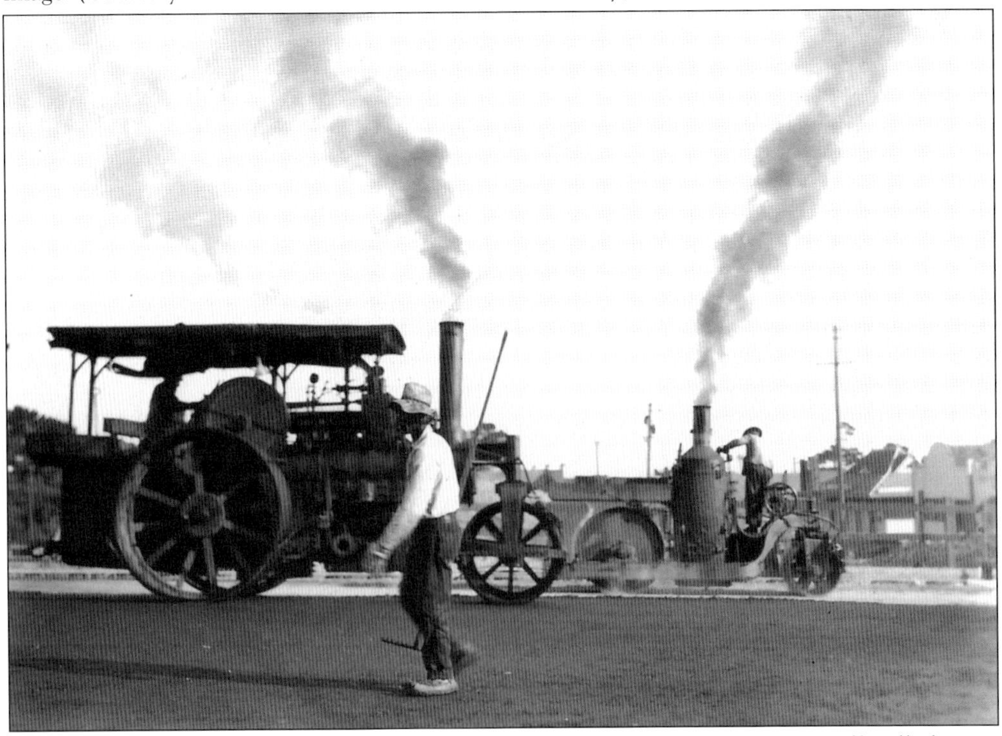

In this C.K. Tuttle collection 1924 photograph, paving of Lighthouse Avenue officially began. Steamrollers can be seen in action in this photograph as they work along the road. Pacific Biological Laboratory, owned by scientist Ed Rickets, can be seen in the image, as well as the browed window of the Pacific Improvement Company office. A worker in the foreground is diligently raking paving. (Courtesy Pacific Grove Museum of Natural History.)

Dark smoke from an oil fire on Cannery Row is visible from Lighthouse Avenue in this photograph taken in front of C.K. Tuttle's pharmacy in 1924. Less than 20 years earlier, another devastating fire affected the Chinese Fishing Village in Pacific Grove. The fire sent a plume of smoke high over Monterey Bay. (Courtesy Pacific Grove Museum of Natural History.)

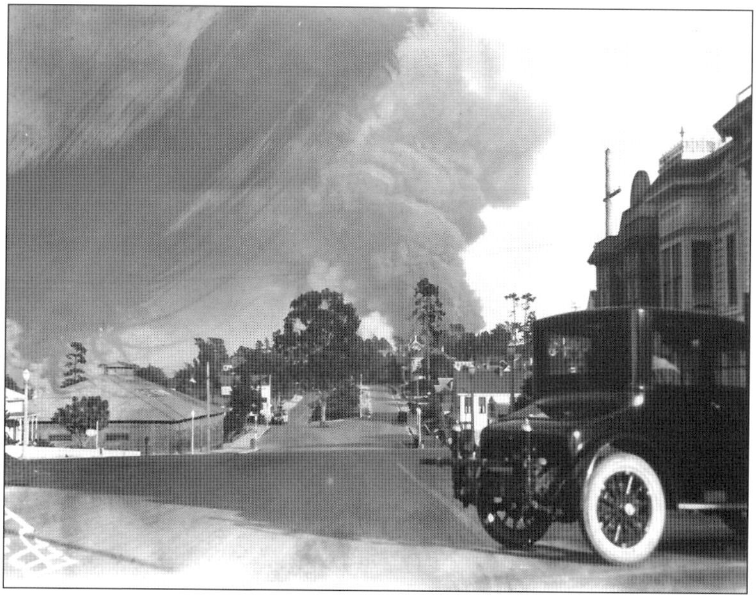

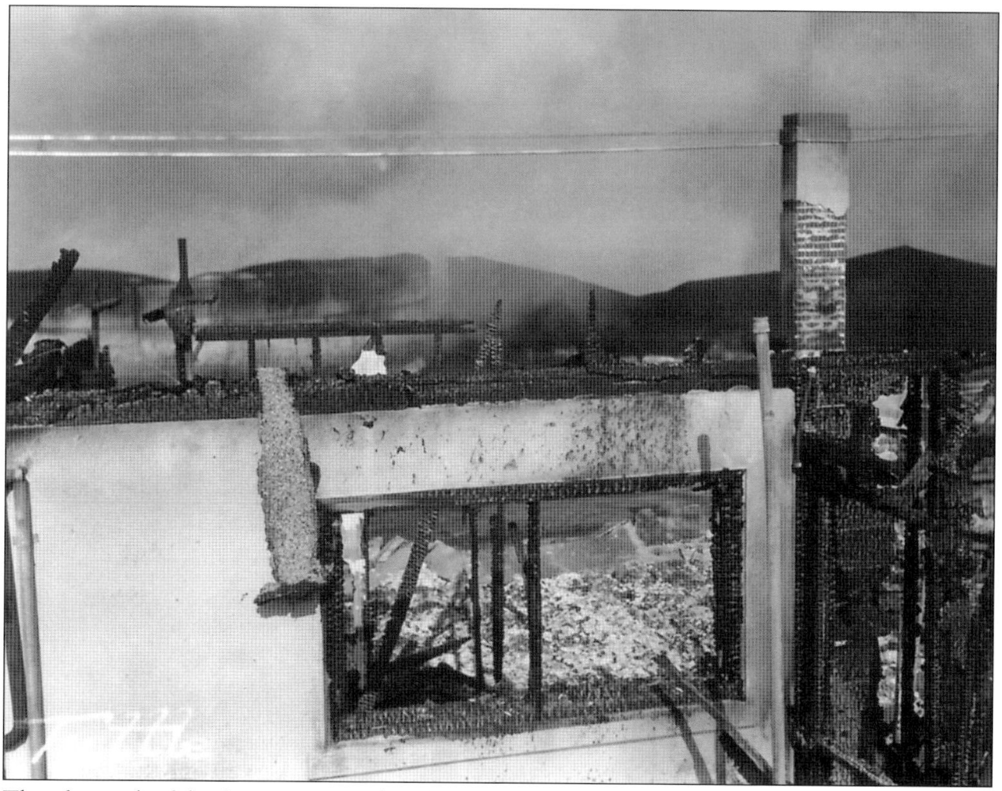

The aftermath of the Cannery Row oil tank fire shows the smoldering remains of a charred building with an empty window visible in the foreground. Modern fire suppression systems are currently in place at the museum to prevent fires like the one that happened in 1924 from happening today. The museum's basement collection is also temperature controlled for preservation purposes. (Courtesy Pacific Grove Museum of Natural History.)

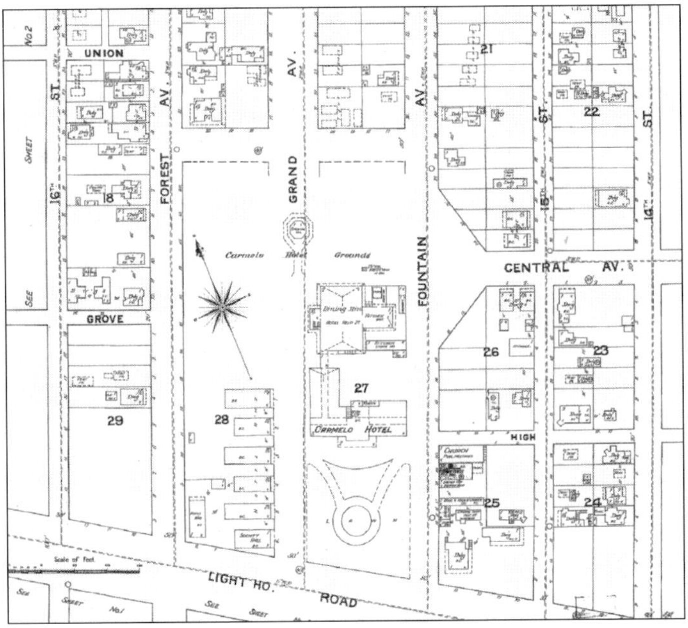

This image shows the museum in 1883 as part of the Sanborn maps. There, the museum is visible in the background, toward the ocean, of the El Carmel Hotel. The map shows the original museum structure on what would today be Grand Avenue before the roads were reconfigured. (Courtesy Pacific Grove Museum of Natural History.)

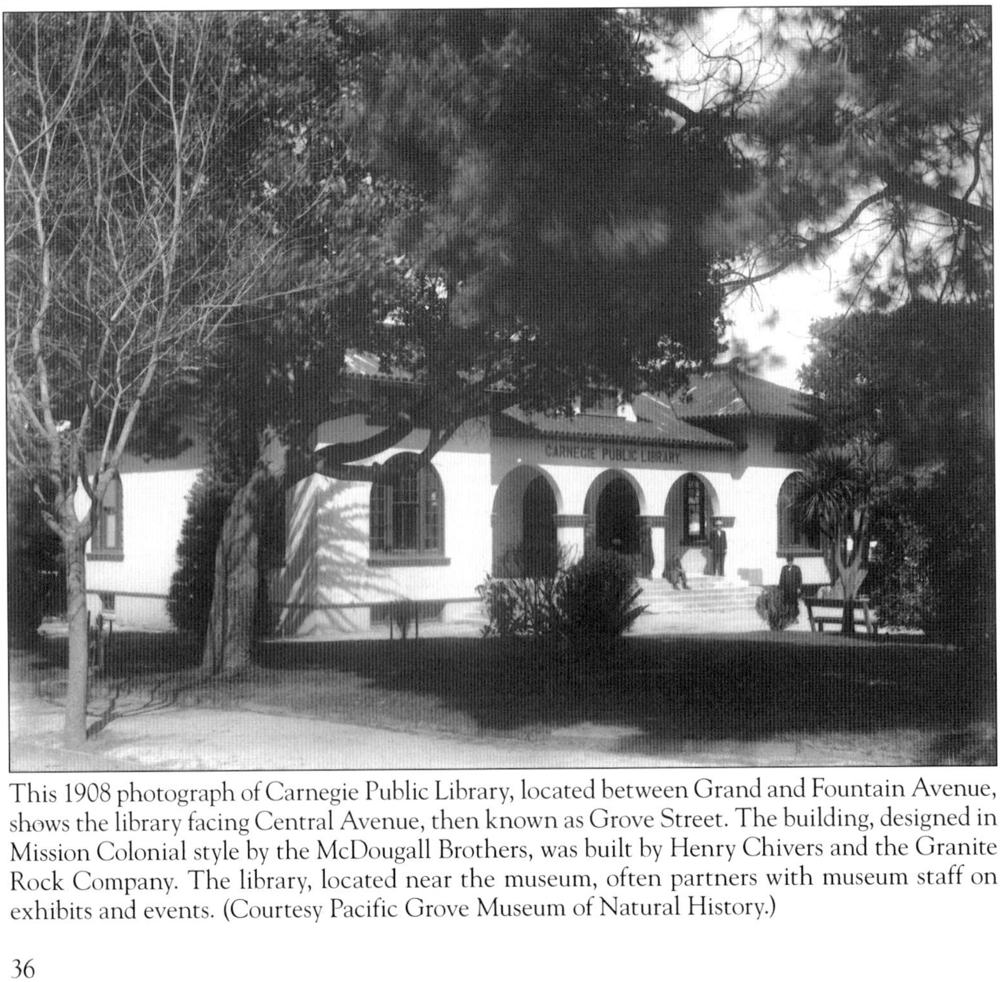

This 1908 photograph of Carnegie Public Library, located between Grand and Fountain Avenue, shows the library facing Central Avenue, then known as Grove Street. The building, designed in Mission Colonial style by the McDougall Brothers, was built by Henry Chivers and the Granite Rock Company. The library, located near the museum, often partners with museum staff on exhibits and events. (Courtesy Pacific Grove Museum of Natural History.)

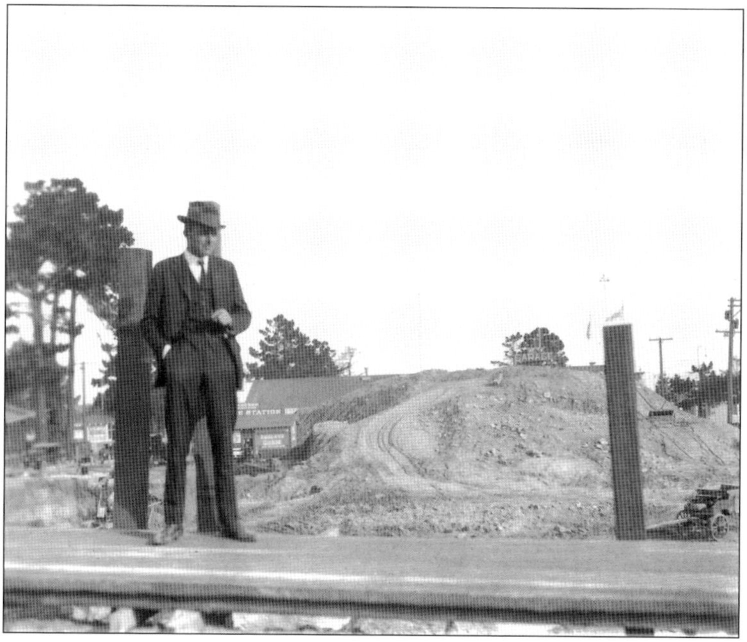

This 1924 C.K. Tuttle image shows a suited man standing before the future site of Holman's Department Store, which would become a neighbor to the museum in later years. Visible in the background is the Associated Gasoline Station and the octagonal chamber of commerce building, a close relative of the original museum building. (Courtesy Pacific Grove Museum of Natural History.)

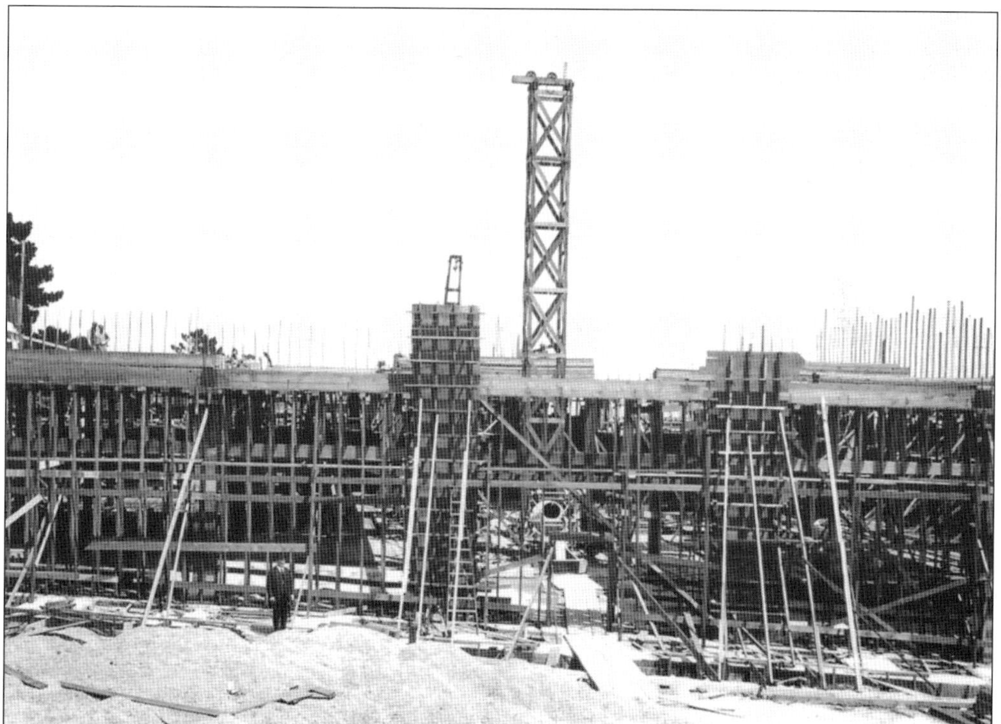

This C.K. Tuttle photograph from 1924 shows construction proceeding on the Holman Department Store with a man in a suit standing at the center of the shot. At this process in the building's construction, excavation for the foundation and the wooden framework is visible. Items from the property currently reside in the museum collection. (Courtesy Pacific Grove Museum of Natural History.)

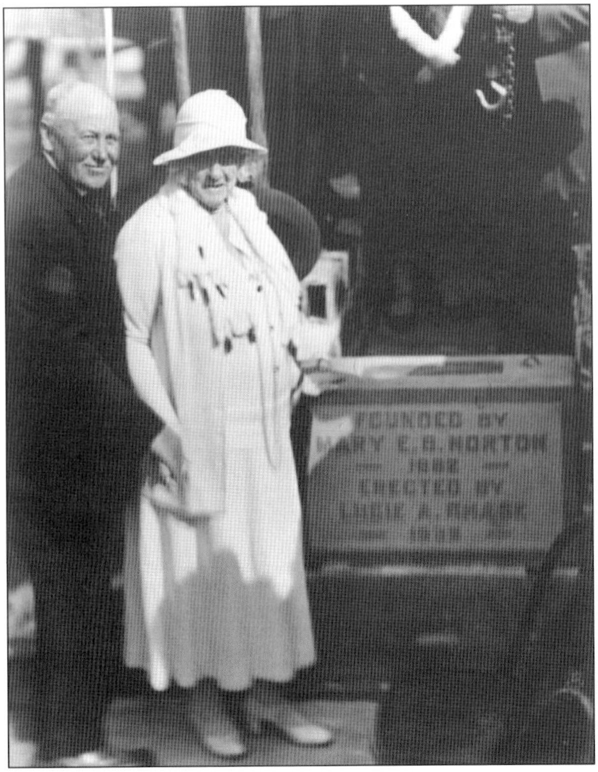

This September 1932 photograph depicts benefactor Lucie A. Chase and Benjamin Lee, Pacific Grove councilman and museum trustee, during the laying of the new museum's cornerstone. Chase was born in 1842 in Plainfield, New Hampshire. She moved to Pacific Grove in 1901 with her husband, Henry. She called laying the cornerstone the happiest day of her life. The new museum's opening ceremony occurred on her 90th birthday. (Courtesy Pacific Grove Museum of Natural History.)

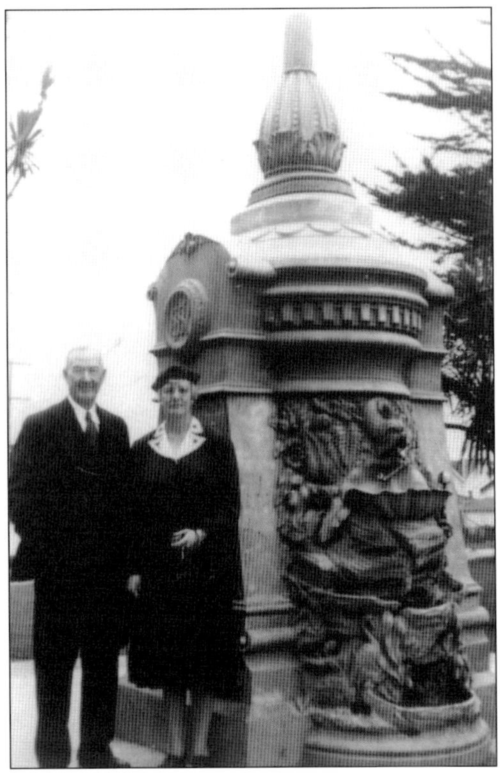

Over the years, Jewell Park continued to be a central gathering location for Pacific Grove residents. In this 1938 photograph, a man and woman can be seen standing next to Cogswell Fountain, which was once located there. The fountain was built to encourage those who passed it to drink water instead of alcohol. The fountain was melted down during a metal drive during World War II. The museum faces Jewell Park today. (Courtesy Pacific Grove Museum of Natural History.)

Three

On Exhibit

Since its inception, the museum has played host to many exhibits and collections, many of which can be found in the large basement located under the building's newest addition. Constructed in 1983, the basement today stores approximately 30,000 singular objects and includes native American baskets, shells, taxidermy, paintings, and much more. These items are often used for temporary exhibits and come from a variety of sources, including museum board member and former scientific director museum director Harold Heath and famed local biologist Ed Ricketts, made famous by John Steinbeck in his novel *Cannery Row* and *The Log from the Sea of Cortez*. Ricketts is thought to have contributed a squid and other unique items that he himself collected from the shores of Monterey Bay. Other notable exhibit items came from ornithologist Rollo Beck. Beck obtained and preserved thousands of specimens for museums worldwide. Hundreds of his specimens can be seen at the Pacific Grove Museum. Some taxidermy specimens still on display at the museum are more than 100 years old. Famed landscape painter Albert T. DeRome is also featured at the museum. His artwork can be seen adorning some of the museum's older exhibits. DeRome's art featured oak woodlands, coastal sand dune landscapes, and much more. In 1915, when the world's fair opened in San Francisco, Monterey County residents were given the opportunity to contribute to the remarkable celebration. On June 11, 1915, the fair, also known as the Panama–Pacific International Exposition, held "Monterey County Day" for visitors to learn more about San Francisco's southern neighbor. This event featured a variety of exhibits displayed during the fair's entirety, which included a chair made from the wood of a tree under which Junipero Serra preached, sardine promotional products, abalone, and a replica of the original custom house. The fair hosted nearly 20 million visitors before it ended. Many of the items on display there found a permanent home at the museum, some of which remain on display to this day.

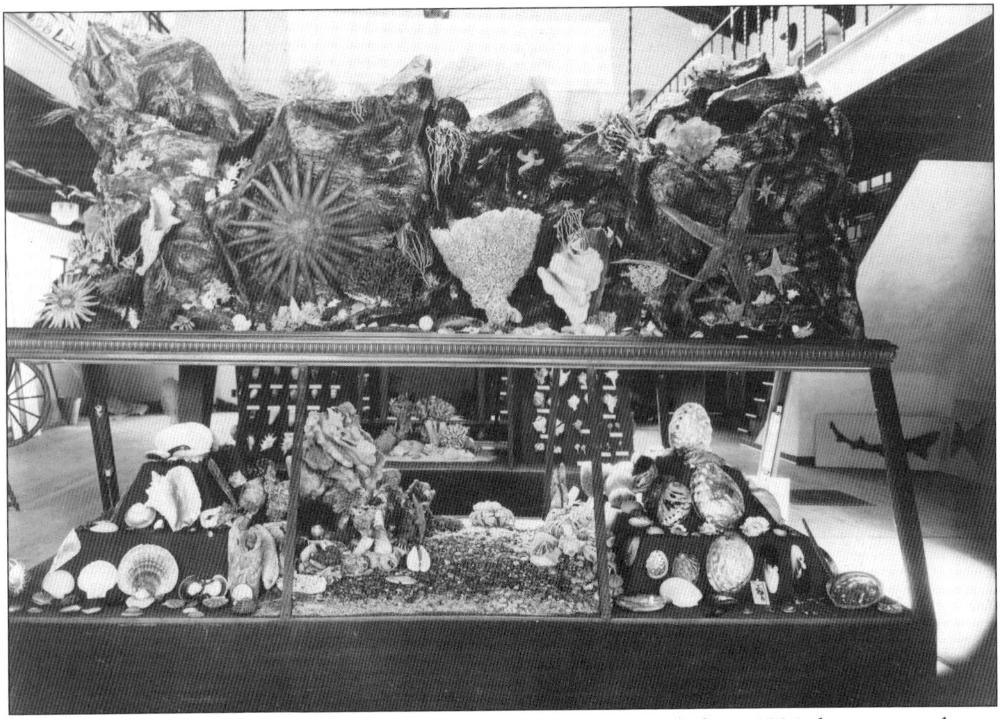

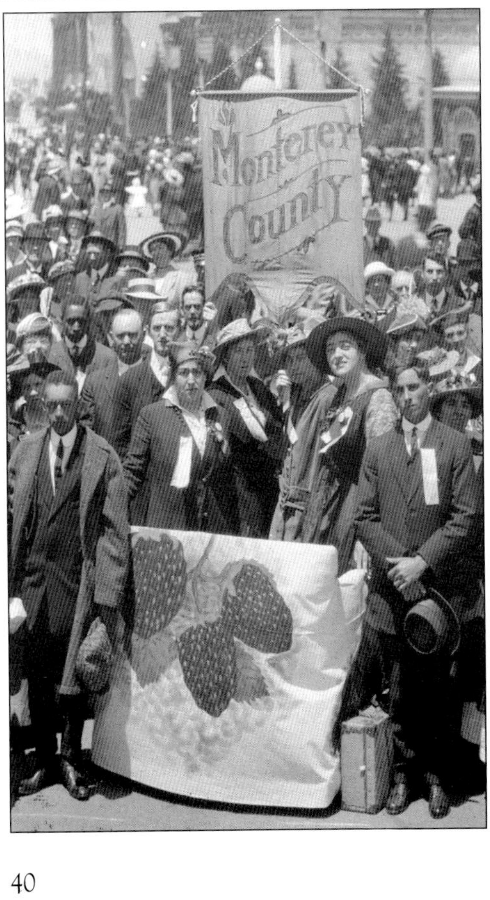

This photograph from 1934 shows an early "dry" aquarium/marine exhibit that was part of the 1915 world's fair in San Francisco: the Panama Pacific International Exposition, which featured a Monterey County Day. The museum in 2015 exhibited many of the same artifacts to celebrate the 100-year anniversary of the event. (Courtesy Pacific Grove Museum of Natural History.)

This photograph from 1915 shows a large group of Monterey County residents as they pose for the camera in San Francisco. The group celebrated Monterey County Day during the Panama Pacific International Exposition. One hundred years later, in 2015, the museum honored the event with a special commemorative exhibit curated by Annie Holdren. (Courtesy Pacific Grove Museum of Natural History.)

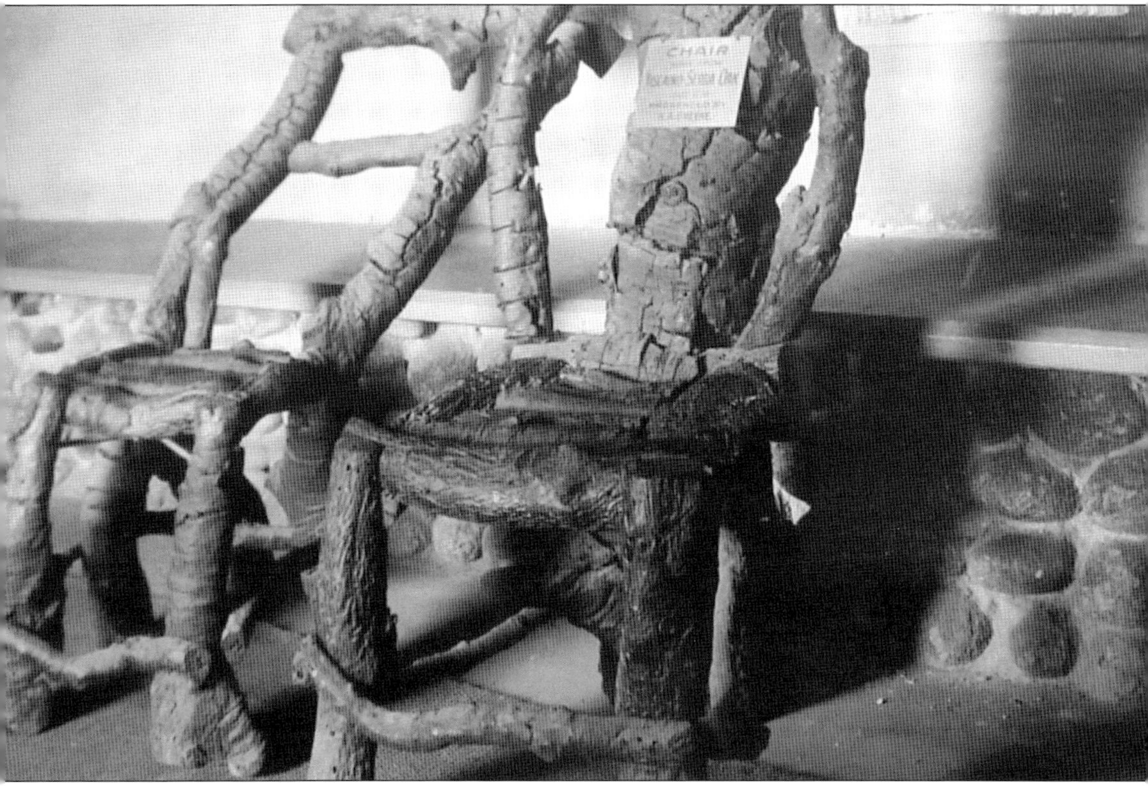

Pictured in 1915 is the chair made from the tree under which St. Junipero Serra held his first sermon after arriving in Monterey County by ship in 1770. The tree survived into the 1900s and, despite being rescued from the sea after it was struck by lightning, was carved up into two chairs, one of which was featured in the 1915 Panama–Pacific International Exposition, otherwise known as the world's fair, in San Francisco as part of a Monterey County–related exhibit. The chair was just one of many intriguing exhibits on display at the time. Junipero Serra was canonized in 2015. The chair was placed on exhibit once again to honor the 100th anniversary of the world's fair in 2015 as well. (Courtesy Pacific Grove Museum of Natural History.)

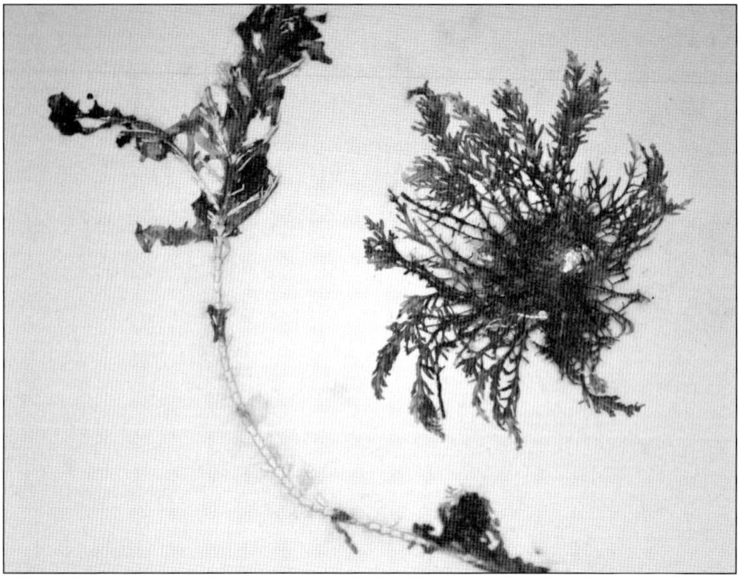

This photograph shows a sea moss specimen collected on July 5, 1892. The specimen was collected by a Mrs. Dewey at Cypress Point in Monterey. The museum's early collection item includes bound books containing hundreds of plant specimens gathered from areas around Monterey County. (Courtesy Pacific Grove Museum of Natural History.)

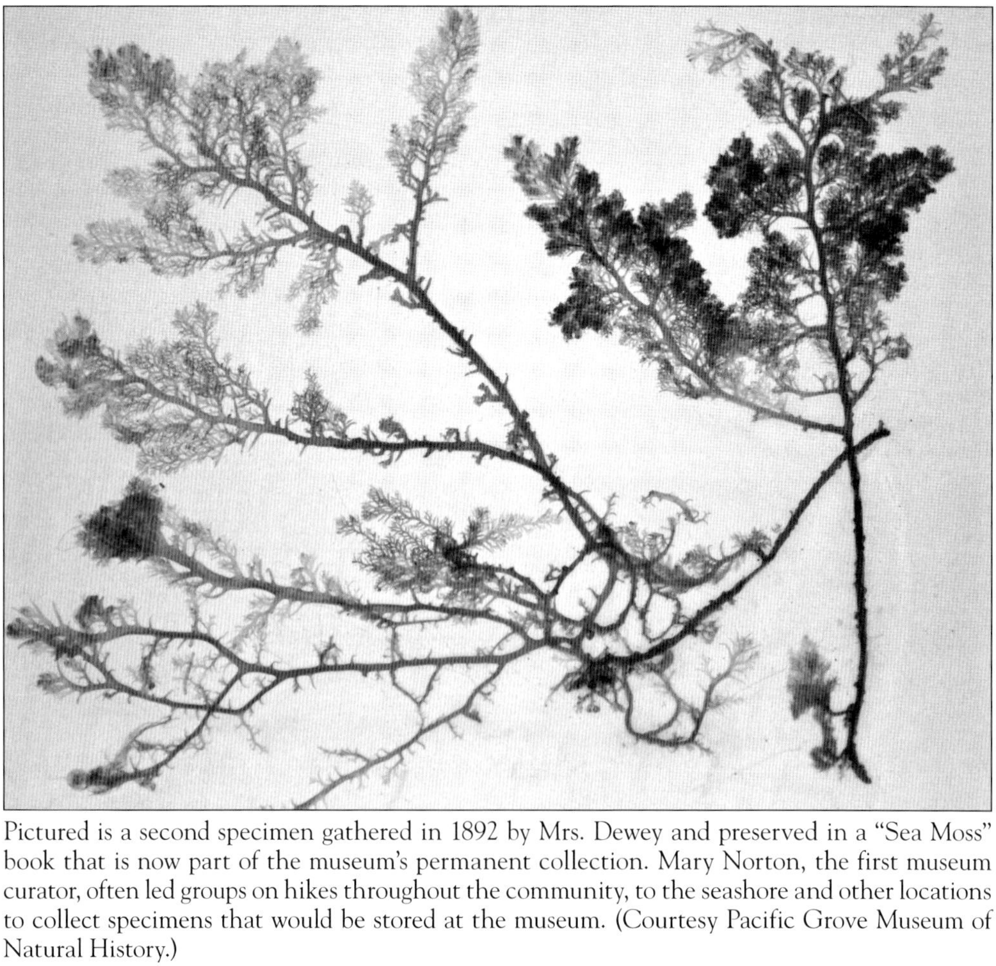

Pictured is a second specimen gathered in 1892 by Mrs. Dewey and preserved in a "Sea Moss" book that is now part of the museum's permanent collection. Mary Norton, the first museum curator, often led groups on hikes throughout the community, to the seashore and other locations to collect specimens that would be stored at the museum. (Courtesy Pacific Grove Museum of Natural History.)

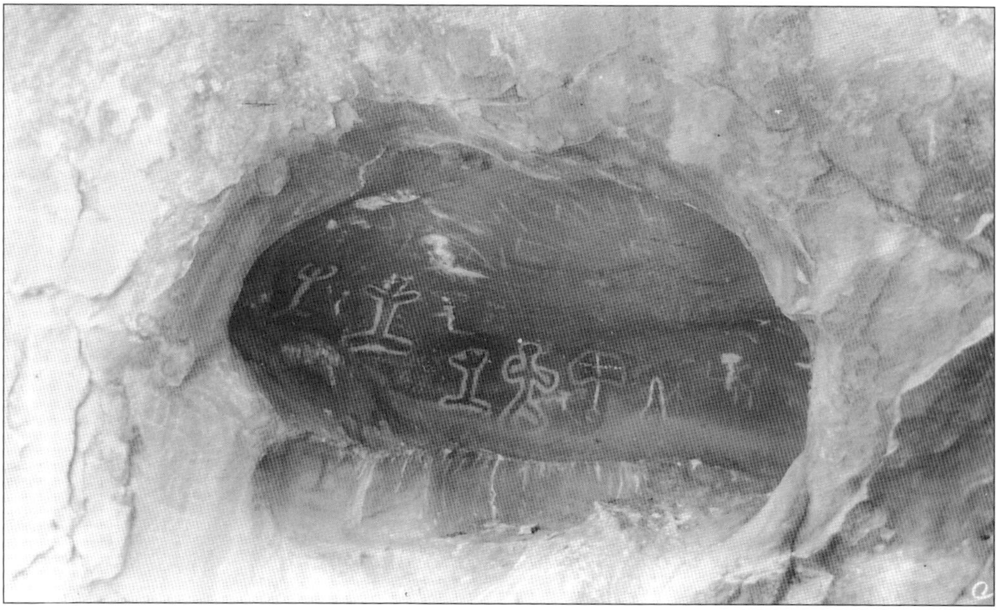

This 1927 photograph shows Native American Chumash rock art from Painted Rock, photographed in the Carrizo Plains in San Luis Obispo County. Human forms can be seen in the rock cavity. Two drawings of the pictographs found on the rock are part of the museum's permanent collection. (Courtesy Pacific Grove Museum of Natural History.)

This 1956 photograph features a Monterey shale specimen with the figure of an eagle carved into the exposed surface, which appears weathered. The appearance of what may be a Mexican eagle as depicted on colonial artifacts from that country is visible on the shale. The object was reportedly collected in Carmel, near Pacific Grove. (Courtesy Pacific Grove Museum of Natural History.)

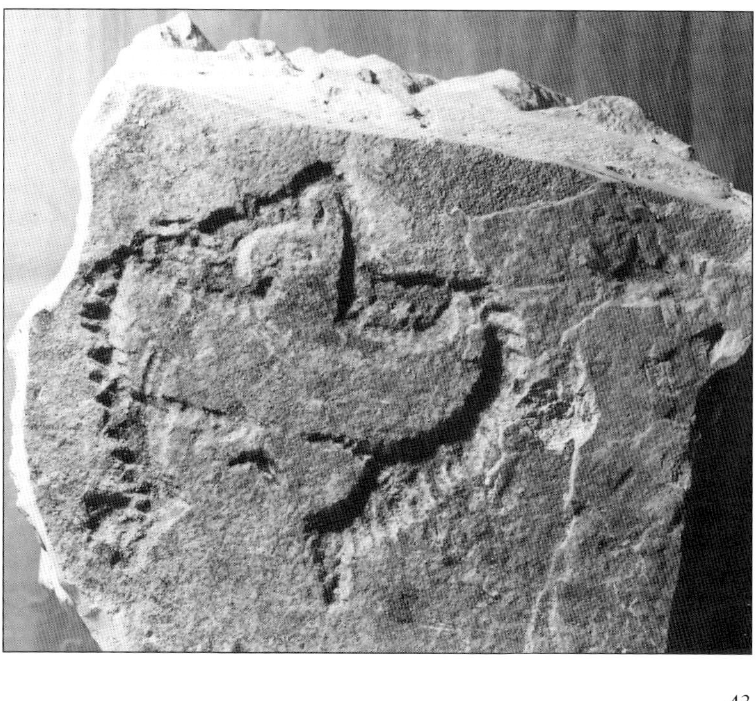

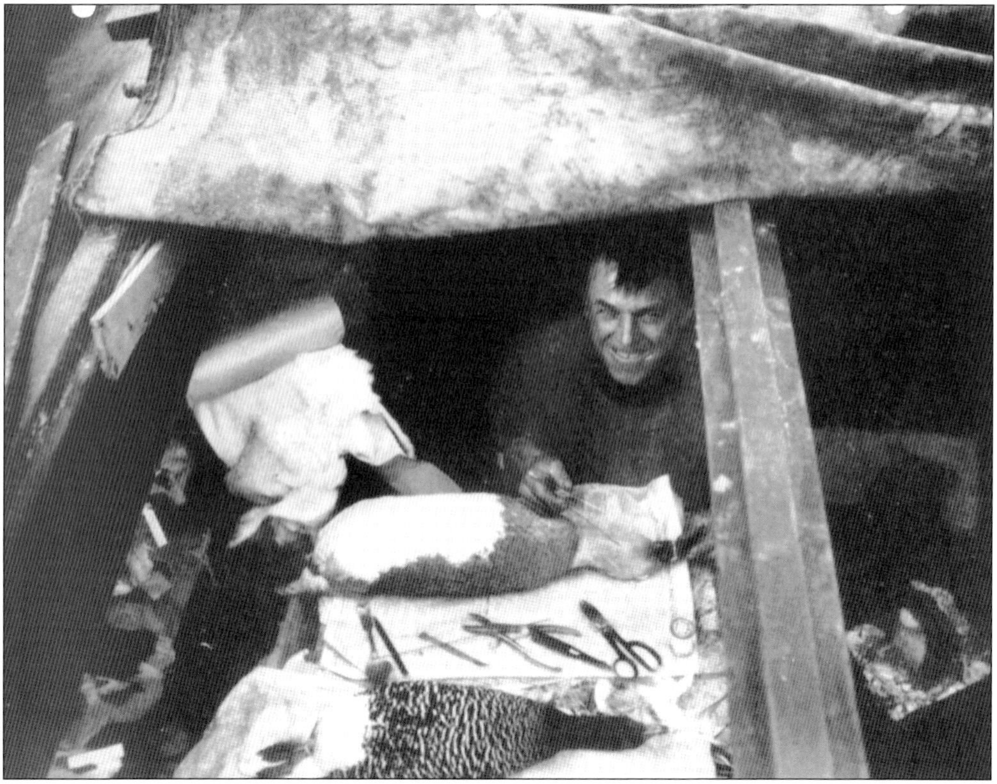

Famous world traveler and ornithologist Rollo Beck can be seen in this early-1900s photograph preparing two bird specimens. His work can be found at several museums, including the natural history museums in England and New York, the California Academy of Sciences in San Francisco, and the museum in Pacific Grove, among other. Beck was born in 1870. (Reproduced from the Library of Congress.)

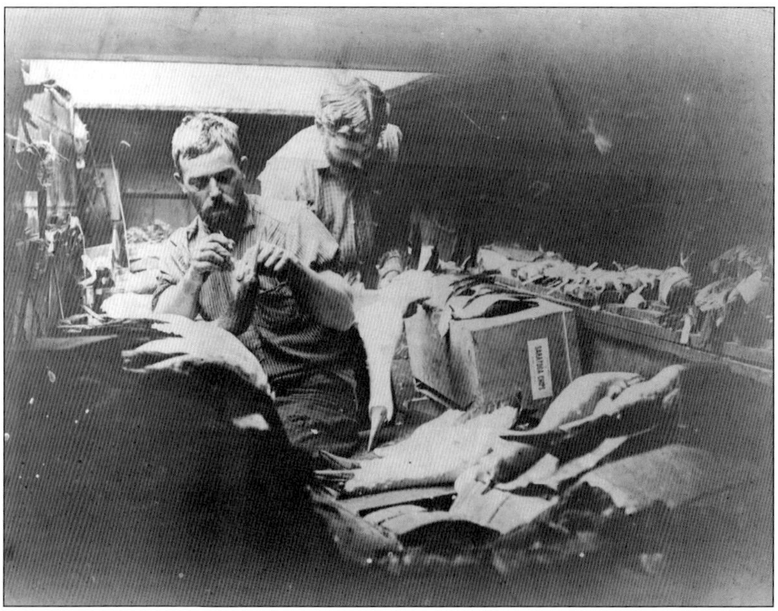

Rollo Beck can be seen in this 1904 photograph preparing a bird specimen for exhibit. Beck contributed to the bird collection at the Pacific Grove Museum of Natural History. In this photograph, Beck is visible, working aboard the *Mary Sachs* ship. Beck died in 1950. (Reproduced from the Library of Congress.)

This Rollo Beck specimen of an adult male osprey (*Pandion haliaetus*) was collected in 1949. Ospreys are not a common sight in the Monterey Peninsula during the summer months, though they have been seen in the area. The museum collection includes more than 400 life-mounted birds. (Courtesy Pacific Grove Museum of Natural History.)

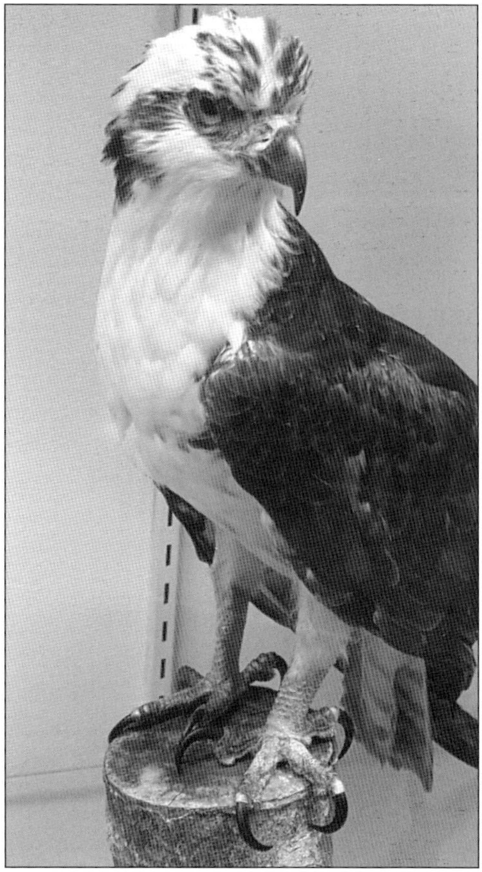

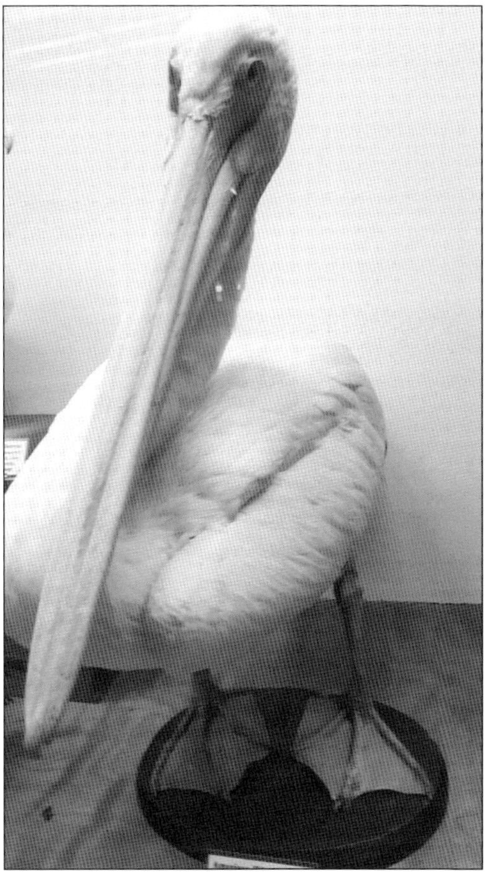

This white pelican was collected in 1932 by Rollo Beck. White pelicans (*Pelecanus erythrorhynchos*) can still be viewed in the area most winters. This pelican is part of the hundreds of local birds found in the museum's bird gallery. All the birds on display there can be seen in the wild throughout Monterey County. (Courtesy Pacific Grove Museum of Natural History.)

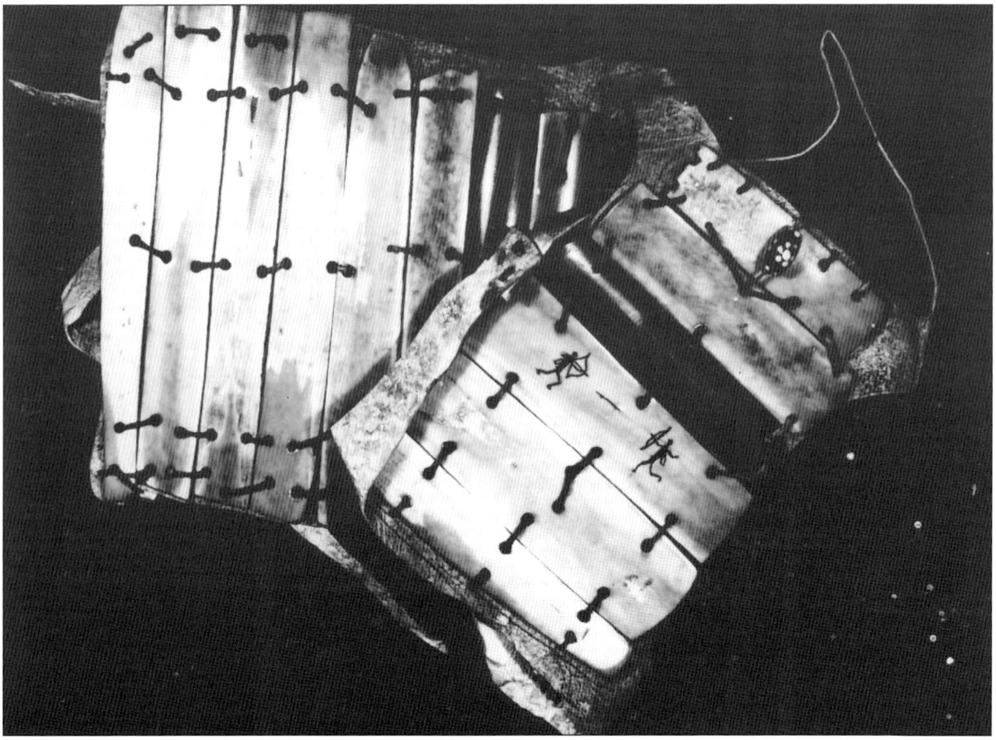

This photograph from 1968 shows a Native American basket gifted to the museum for its collection. Over the years, the museum has featured many basketry exhibits that detail the area's rich history, as well as the history of basket making itself. A Native American baskets exhibit can still be found at the museum today. (Courtesy Pacific Grove Museum of Natural History.)

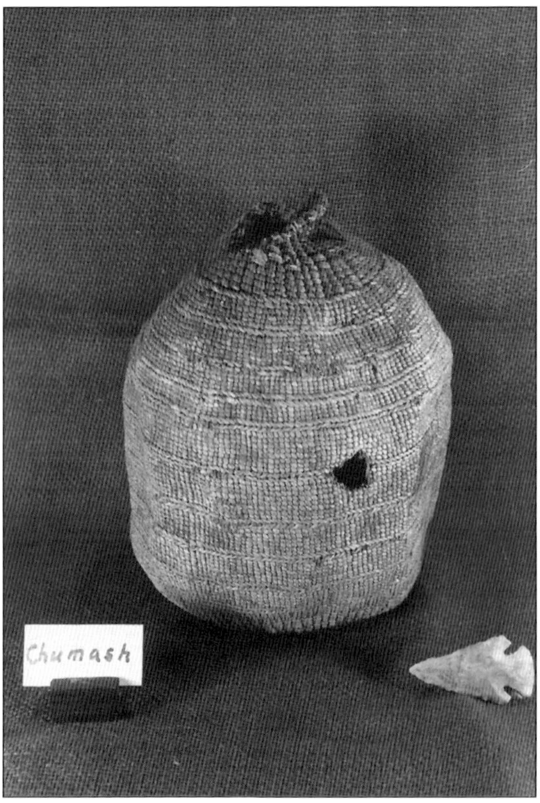

This Chumash twined water bottle is made of tule rushes lined with asphaltum, a thick, sticky tar-like substance. The artifact was collected from San Miguel Island, which is in the Pacific Ocean off the Southern California coast near Los Angeles. (Courtesy Pacific Grove Museum of Natural History.)

Taken from the museum's basement collection, this is an Owens Valley Paiute doll cradle, beautifully made with yarn and cloth trimmed. Records for the item indicate it was collected in the state of Nevada. The date it was collected is unknown. (Courtesy Pacific Grove Museum of Natural History.)

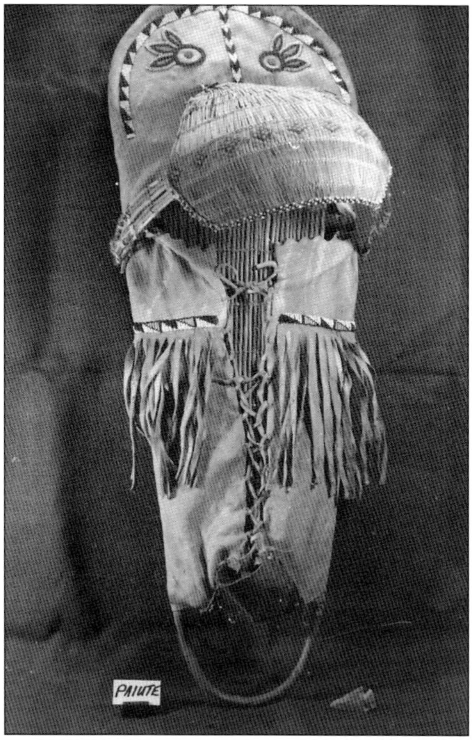

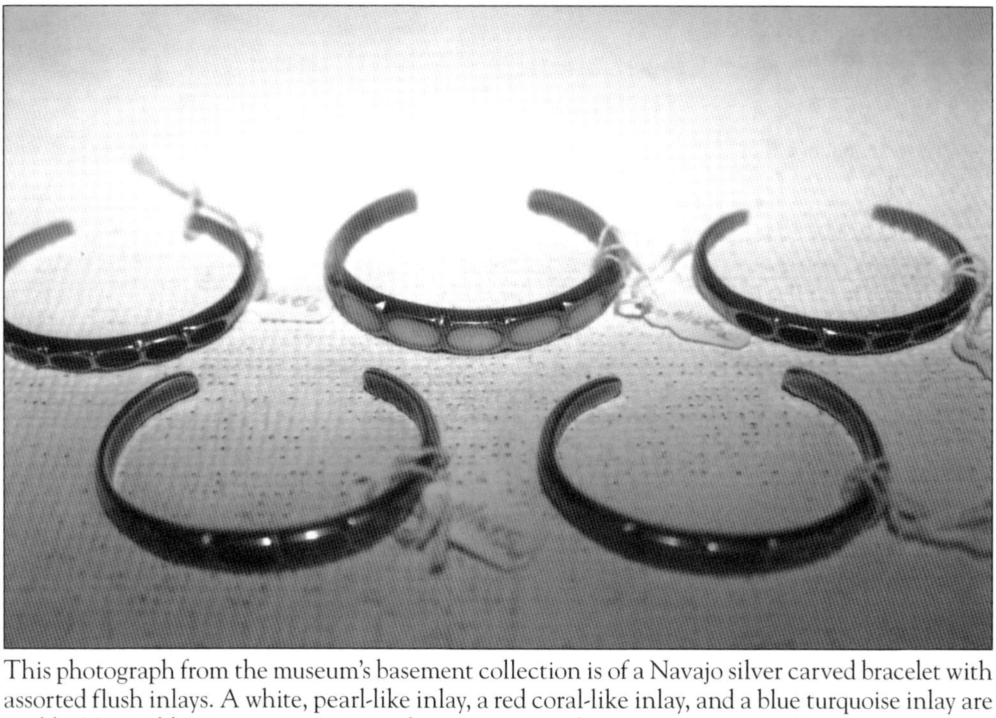

This photograph from the museum's basement collection is of a Navajo silver carved bracelet with assorted flush inlays. A white, pearl-like inlay, a red coral-like inlay, and a blue turquoise inlay are visible. Natural history museums are often recipients of items. Sometimes, the items are turned away, and other times, the items are made official collection items through the accessioning process. (Courtesy Pacific Grove Museum of Natural History.)

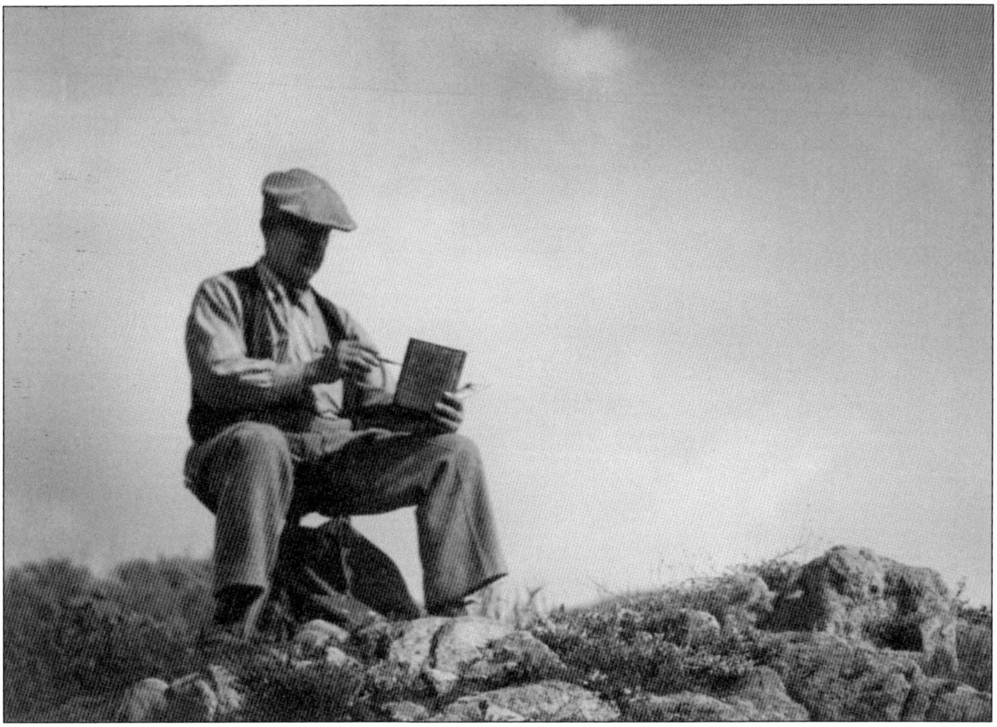

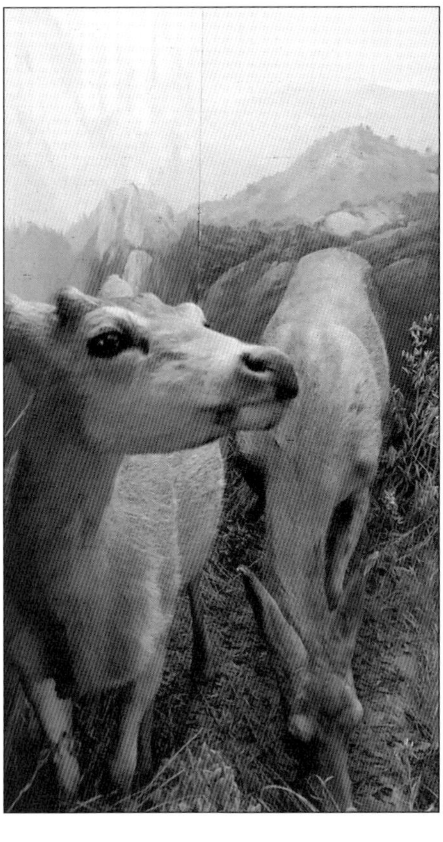

Famed landscape artist Albert DeRome, seen in this self-portrait from 1955, could often be found painting along the coast. DeRome, a native of California, studied at the Mark Hopkins Institute of Art in San Francisco. He painted many of the backgrounds for the museum's diorama cases and remained a resident of Pacific Grove until his death in 1959. (Courtesy Pacific Grove Museum of Natural History.)

DeRome painted many background landscapes used by the museum in its exhibits, including the deer case exhibit in the 1950s, pictured. The case can be found today in the Heritage Gallery at the museum. DeRome's work can still be seen throughout the museum. Several of his paintings can be found in the museum's basement collection. (Courtesy Pacific Grove Museum of Natural History.)

Many in Pacific Grove bring their personal collection items or belongings collected by a family member to the museum to accession into the permanent collection. Often, those pieces can come from exotic, faraway locales. Pictured here is pottery collected from Panama in 1956. (Courtesy Pacific Grove Museum of Natural History.)

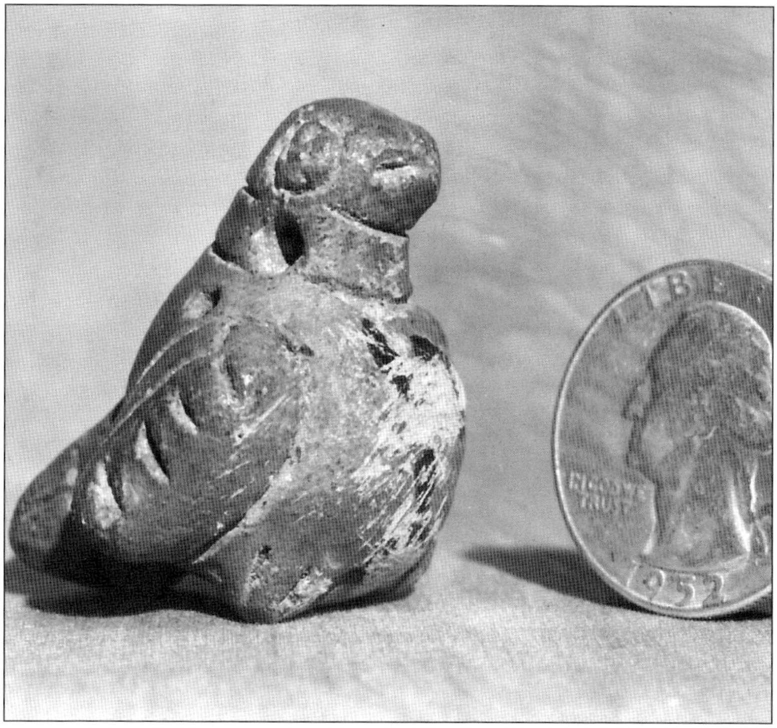

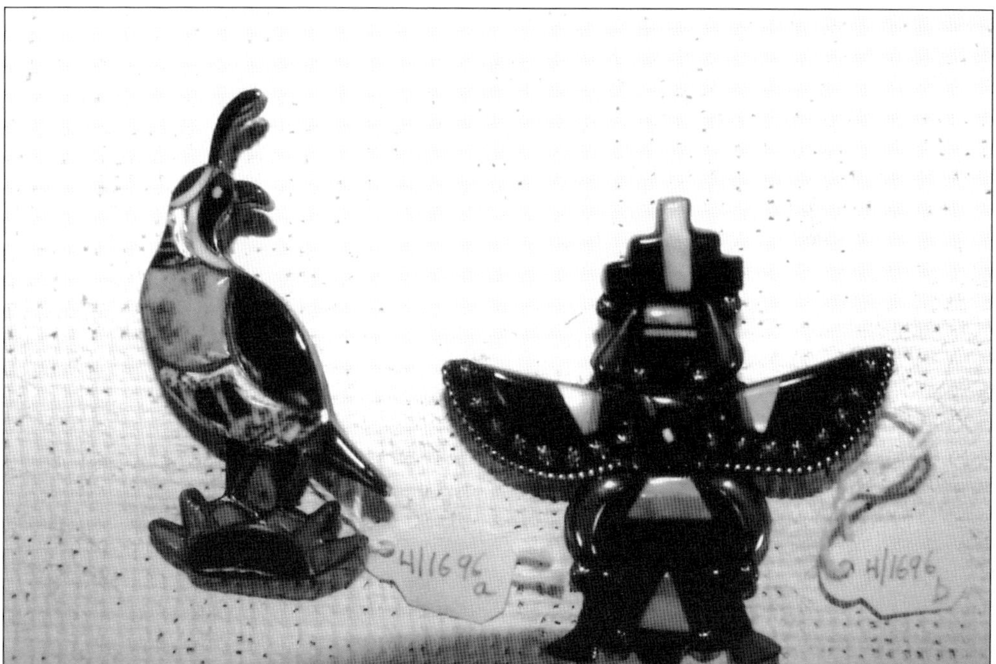

This image from the museum's basement collection features Navajo decorative pins complete with a multicolored inlay. At left is a quail standing on a rising sun, and at right is a kachina doll/Kwahu kachina, or eagle. The date of the photograph is unknown. (Courtesy Pacific Grove Museum of Natural History.)

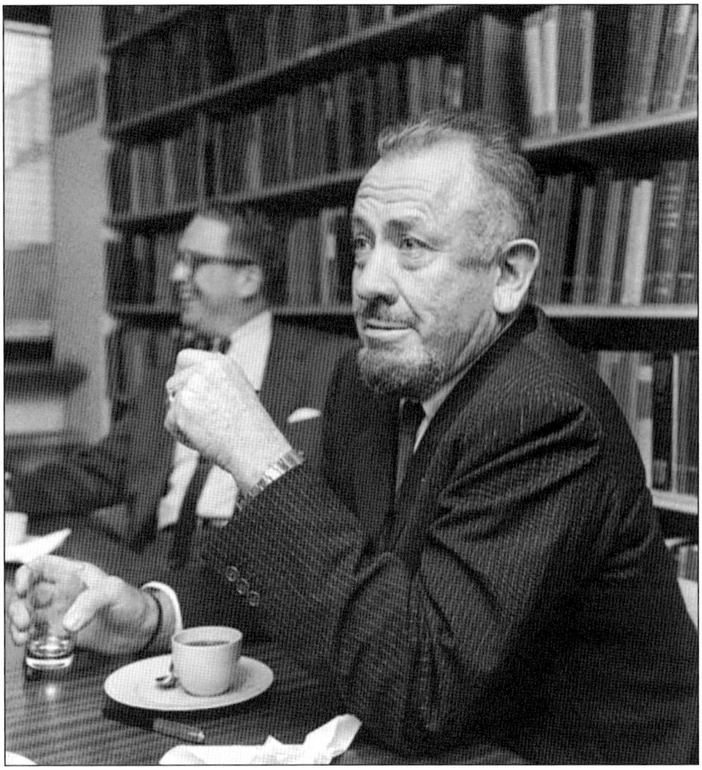

Pictured is author and Nobel Prize winner John Steinbeck, who was a resident of Pacific Grove for some time. He is pictured here in Finland in October 1963, a year after receiving the prize. Steinbeck was a good friend of Ed Ricketts, who collected marine specimens along the Monterey County shoreline. Steinbeck wrote about his friend in the books *Cannery Row* and *The Sea of Cortez* as well as other publications under different names. (Photograph by Olavi Kaskisuo/Lehtikuva.)

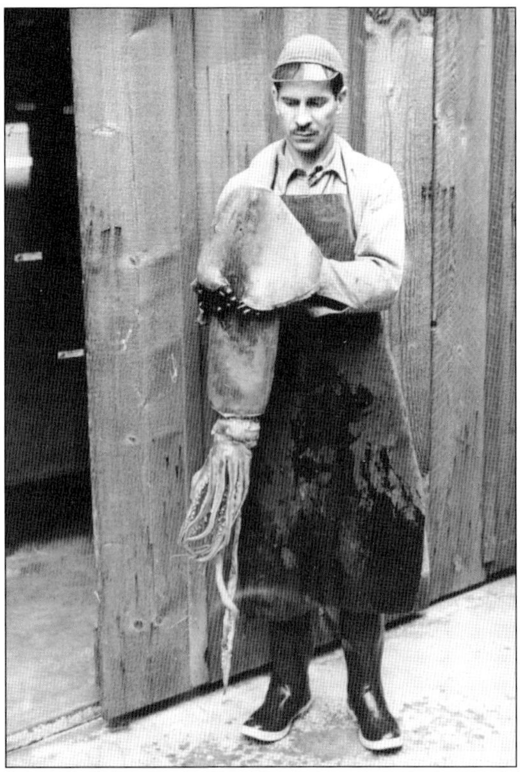

This 1935 photograph by Ralph Buchsbaum, used with permission from Vicki Buchsbaum Pearse, shows marine naturalist and philosopher Ed Ricketts holding a jumbo squid (*Dosidicus gigas*) prior to preparing the amazing specimen for exhibit. Ricketts contributed several specimens to universities and museums across the nation, including the museum in Pacific Grove. (Photograph by Ralph Buchsbaum, courtesy Vicki Buchsbaum Pearse.)

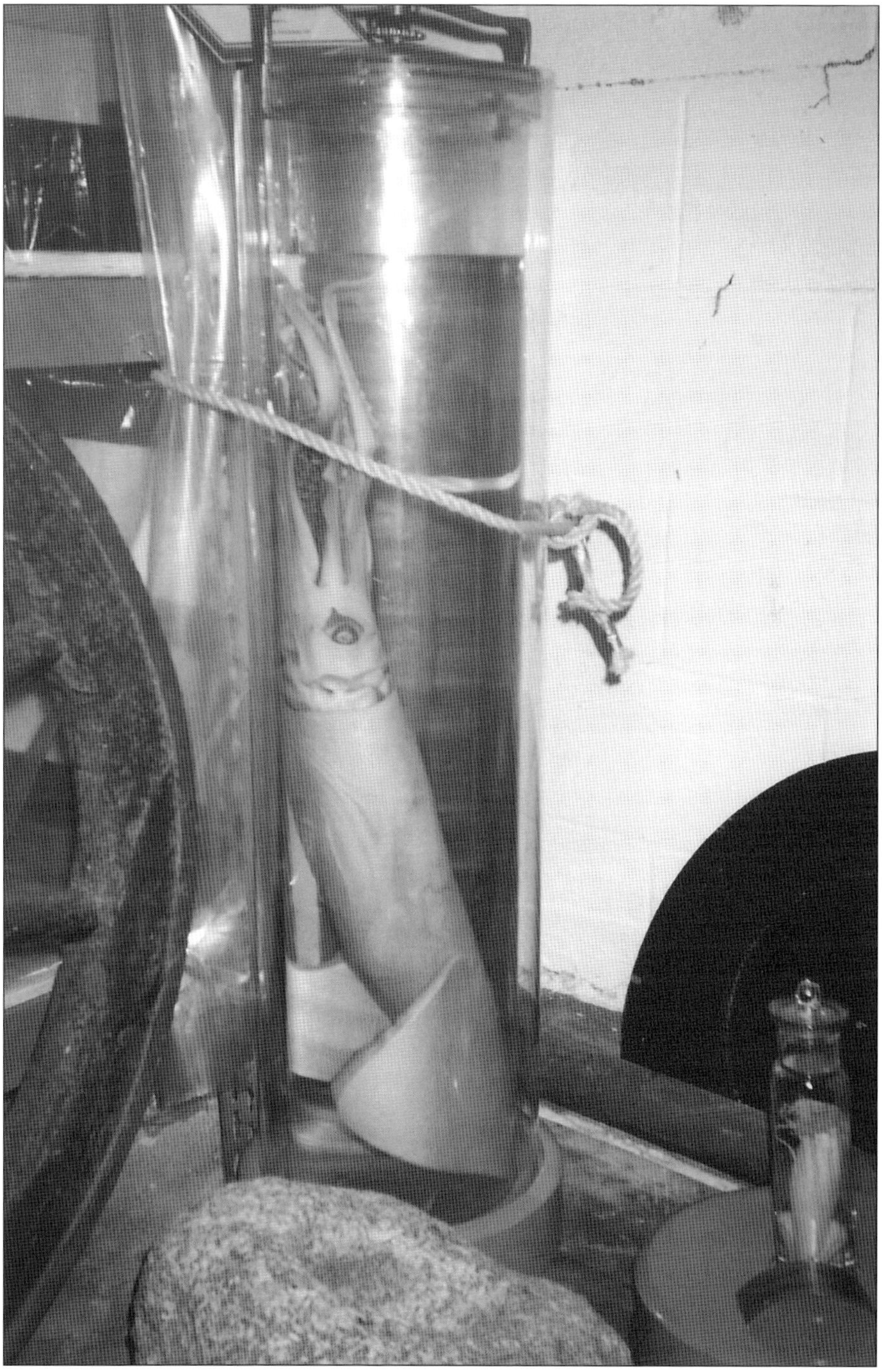

Pictured is a *Dosidicus gigas*, otherwise known as a Humboldt squid, which was collected in 1935 by Ed Ricketts, according to museum records. The squid is currently on display at the museum as part of an exhibit honoring important collectors throughout the museum's long history. Ricketts, a famed local scientist at the time, is part of that exhibit. A smaller squid can also be seen to the right of the larger one in this image. A second squid of the same size can be found in the museum lobby as a donation location. (Courtesy Pacific Grove Museum of Natural History.)

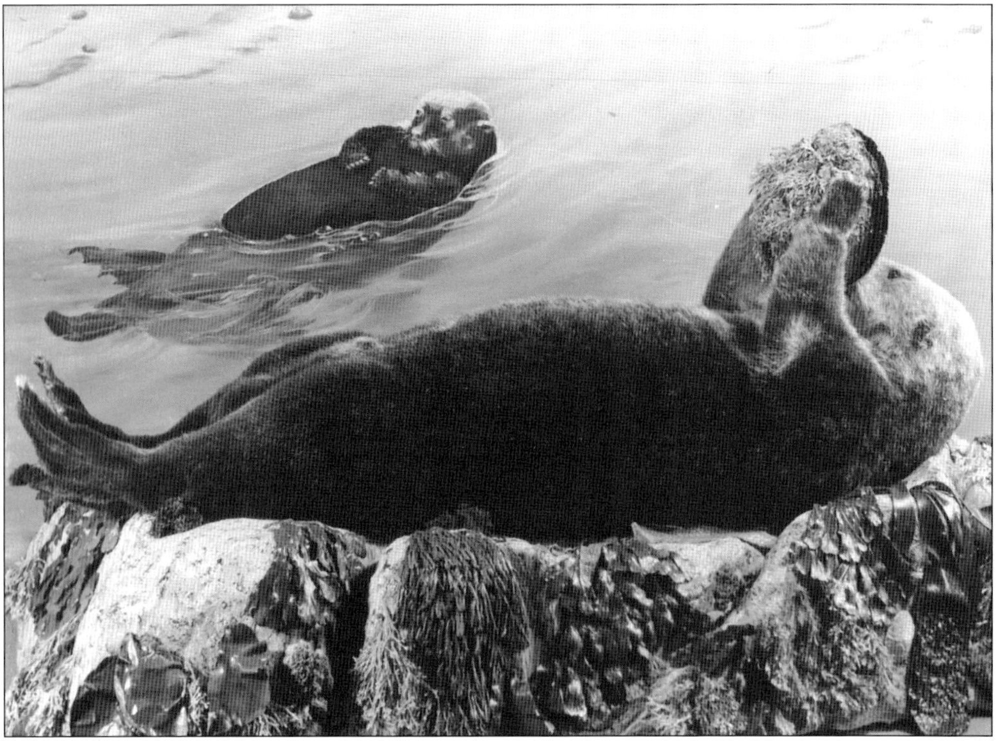

This adult male sea otter was collected in 1948 off the shoreline near Big Sur, according to records. The otter was made into a taxidermy specimen in 1950 in Salinas and then later accessioned into the museum's collection and put on exhibit, where it remains to this day, though its location within the museum has changed over the years. (Courtesy Pacific Grove Museum of Natural History.)

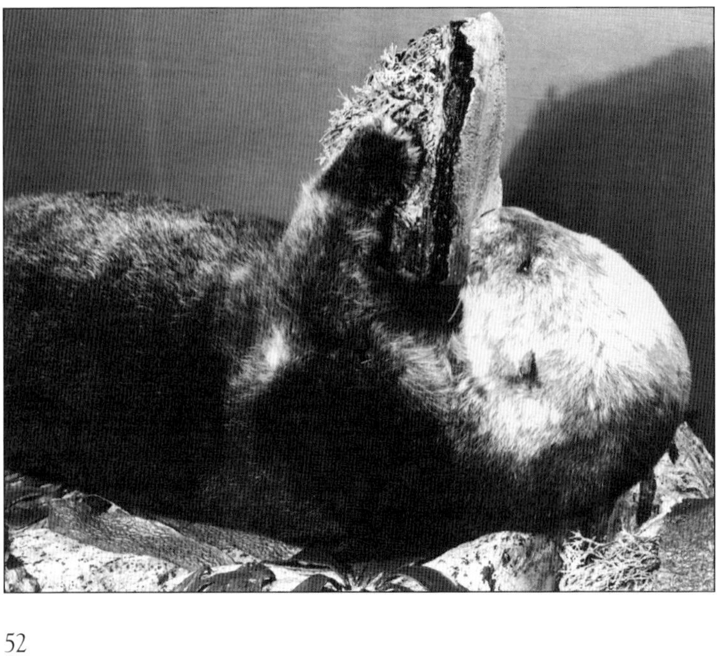

This is another view of the adult male sea otter collected by the museum in 1950. The otter was taken off display briefly in 1991 due to glass breakage, according to museum records. The exhibit was later fumigated in 1991 and taken off exhibit at least once more before it took its current location on the museum's mezzanine level. (Courtesy Pacific Grove Museum of Natural History.)

Mammals are often prepared to be placed on exhibit and accessioned into the museum's collection, from the museum's earliest days and onward. The practice of taxidermy, with some changes, remains consistent, as evidenced by this female Merriam's chipmunk (*Tamias merriami*), which was prepared in 1990 and displayed as part of the museum's Mammalia exhibit. (Courtesy Pacific Grove Museum of Natural History.)

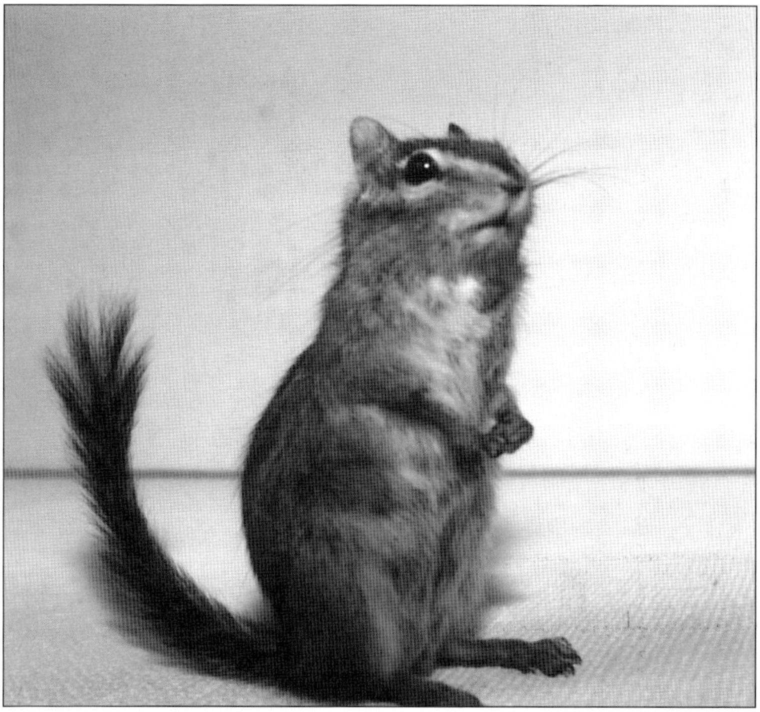

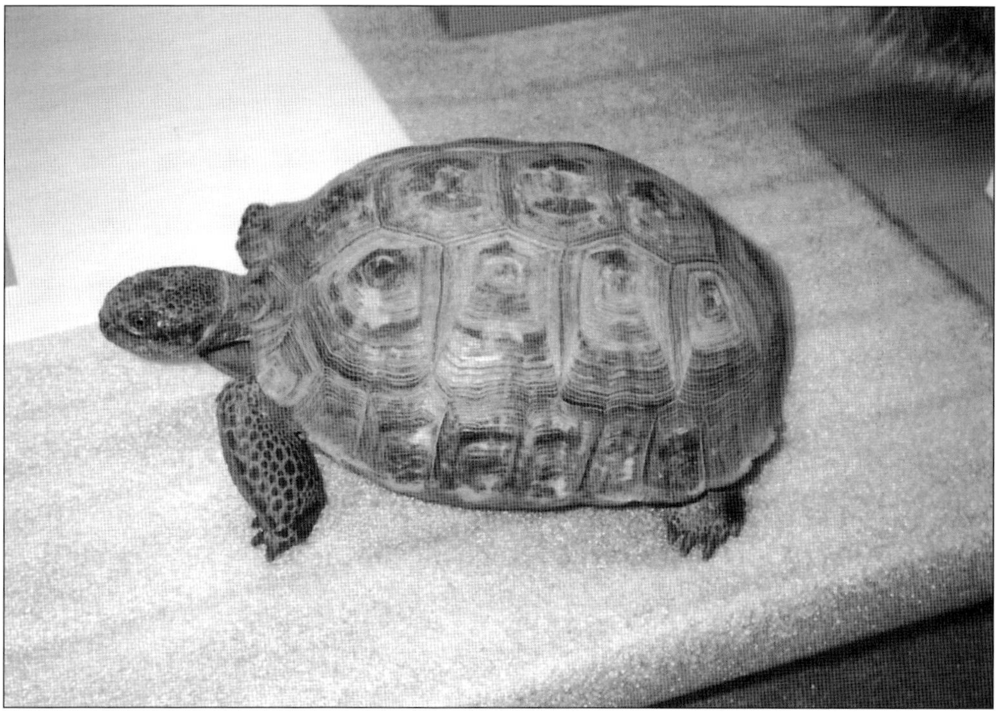

This handsome tortoise from the basement collection, seen here in this photograph from the early 1980s, is an example of collection items that are given a temporary catalogue number. This identifies the object in the museum's database for general information and to establish provenance, which clarifies the object's history. (Courtesy Pacific Grove Museum of Natural History.)

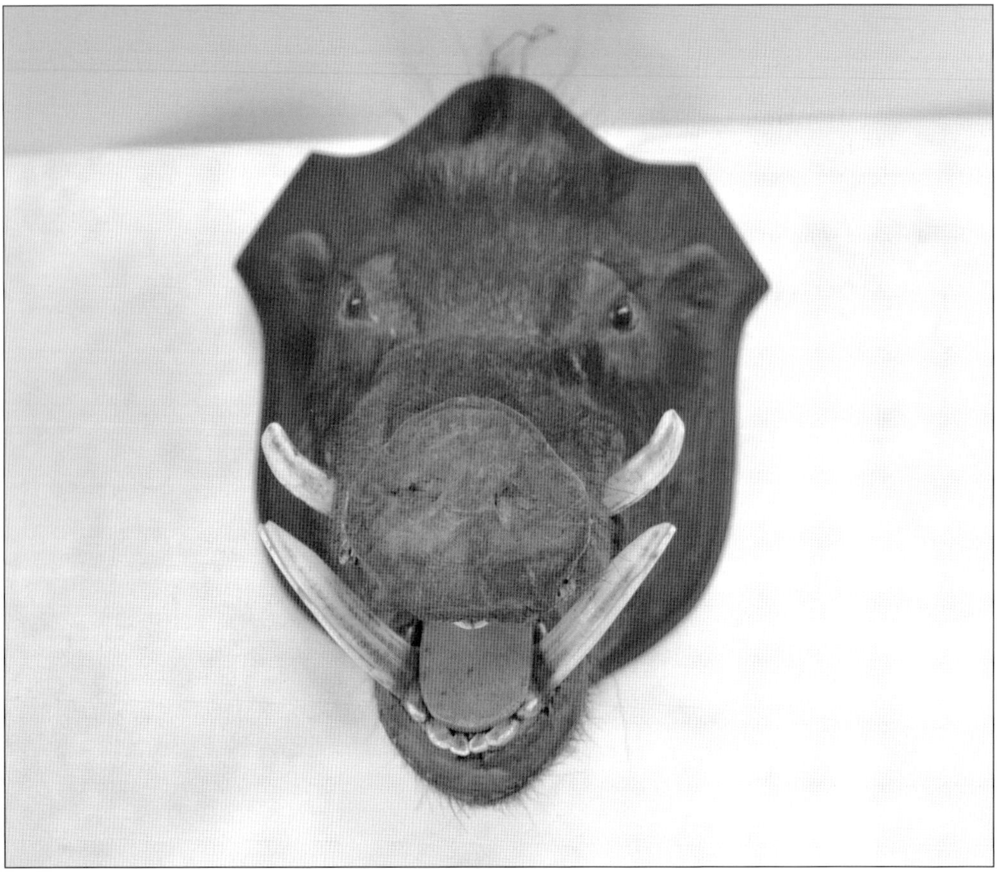

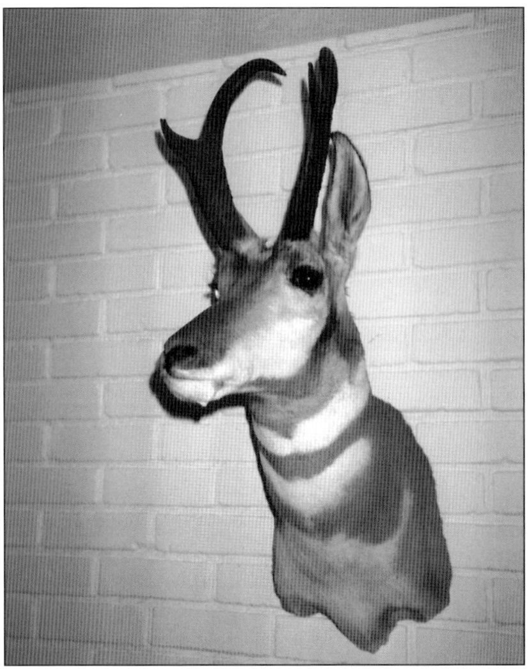

The museum has an extensive collection of mounted taxidermy animal heads, including this warthog, collected by Henry Edward Odell in January 1938. Odell, a veteran of the Navy, collected the specimen while overseas. He eventually came to call the community of Carmel, located close to Pacific Grove, his home. (Courtesy Pacific Grove Museum of Natural History.)

While a common sight in the southwestern states, pronghorn antelope, such as this one gifted to the museum in 1988, are not a particularly common sight along the Central Coast. The museum collection consists of a number of other taxidermy mounted heads, including a seal and a boar. (Courtesy Pacific Grove Museum of Natural History.)

Pictured is a Ho'o'te kachina that was displayed in the 1960s as part of the Kachina Dolls: Hopi Spirit exhibit at the museum. The dance of this kachina forecasts the flowers of spring, according to the curator's notes documented at the time of the exhibit. This kachina doll was made from wood and ornately decorated. (Courtesy Pacific Grove Museum of Natural History.)

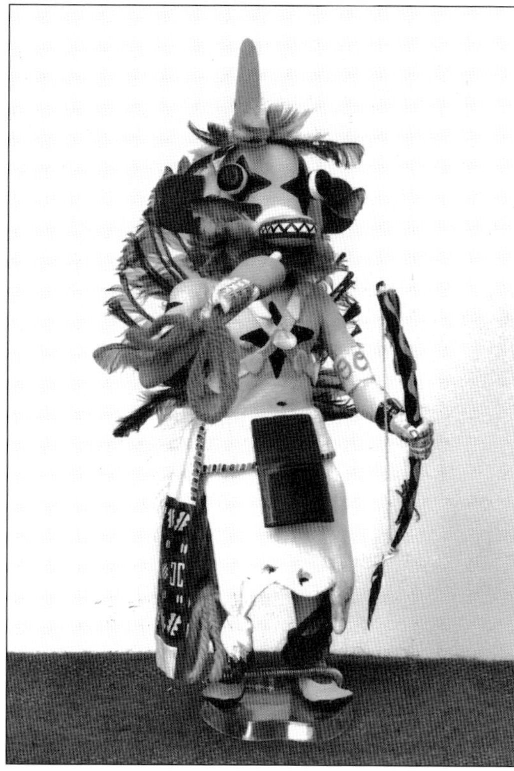

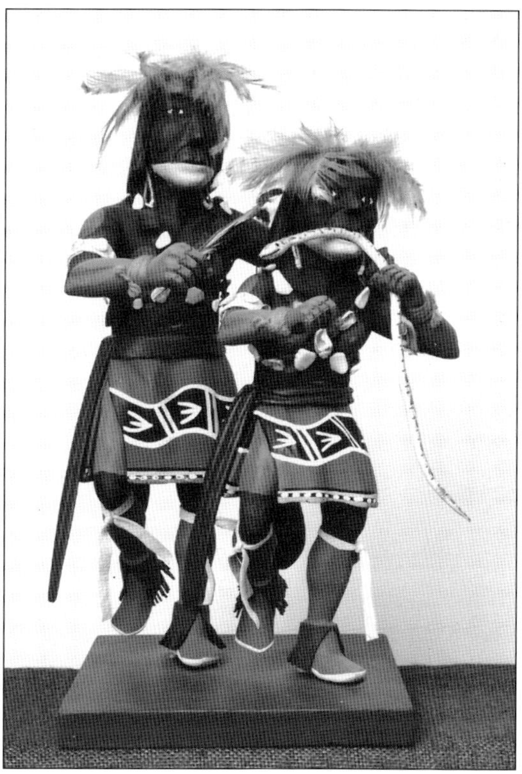

Seen here is another Hopi Spirit kachina that depicts two figures, one with a snake in its mouth. The figures were likely used, along with other kachinas such as this, in the Room of Man to display information about Native Americans. This exhibit featured items from others and did not focus entirely on the Central Coast. (Courtesy Pacific Grove Museum of Natural History.)

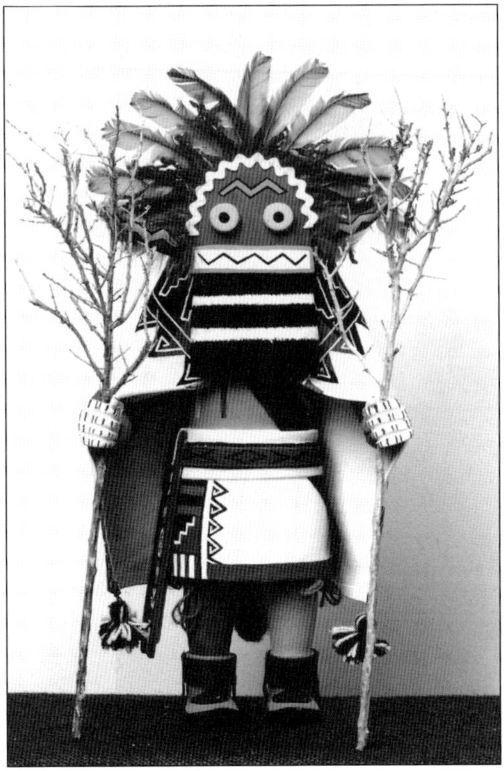

Pictured is Wuyak-kuita, the broad-faced Hopi Spirit kachina. This photograph from the 1960s was part of several exhibits held at the museum through the 1980s. This kachina was made from cottonwood wood root that was painted and decorated. kachina dolls were typically made from cottonwood root. (Courtesy Pacific Grove Museum of Natural History.)

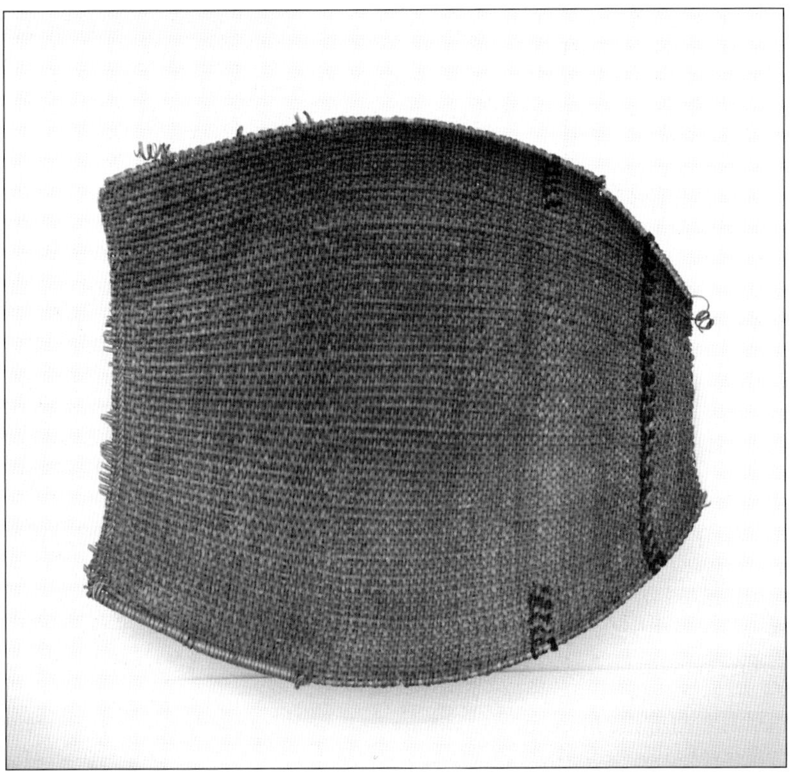

This photograph from 1976 shows a California Indian basket that was part of a Native American basketry exhibit that year. The basket is said to be a Costanoan Winnowing Tray, or "Walahin." A Native American basket exhibit can currently be found on the second floor of the museum. (Courtesy Pacific Grove Museum of Natural History.)

This photograph from 1976 offers a close-up view of a Native American basket with an equisetum root decoration (horsetail), used for marketing the California Indian basket show to residents and visitors to Monterey County. The beautiful image offers an intriguing view of the detail involved in creating a basket. (Courtesy Pacific Grove Museum of Natural History.)

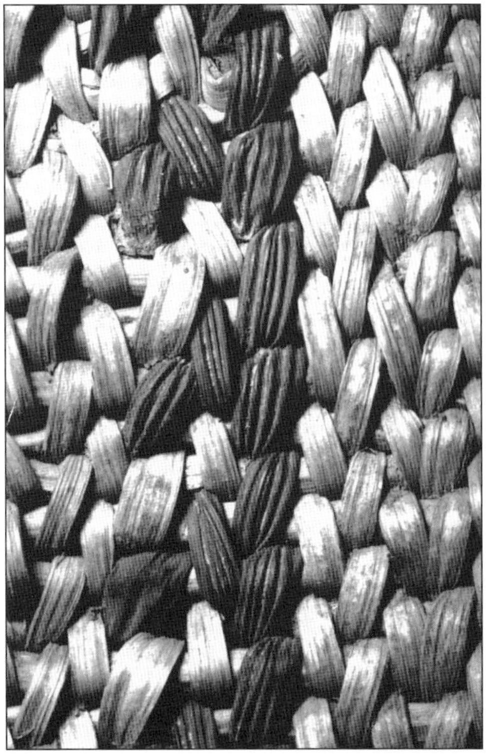

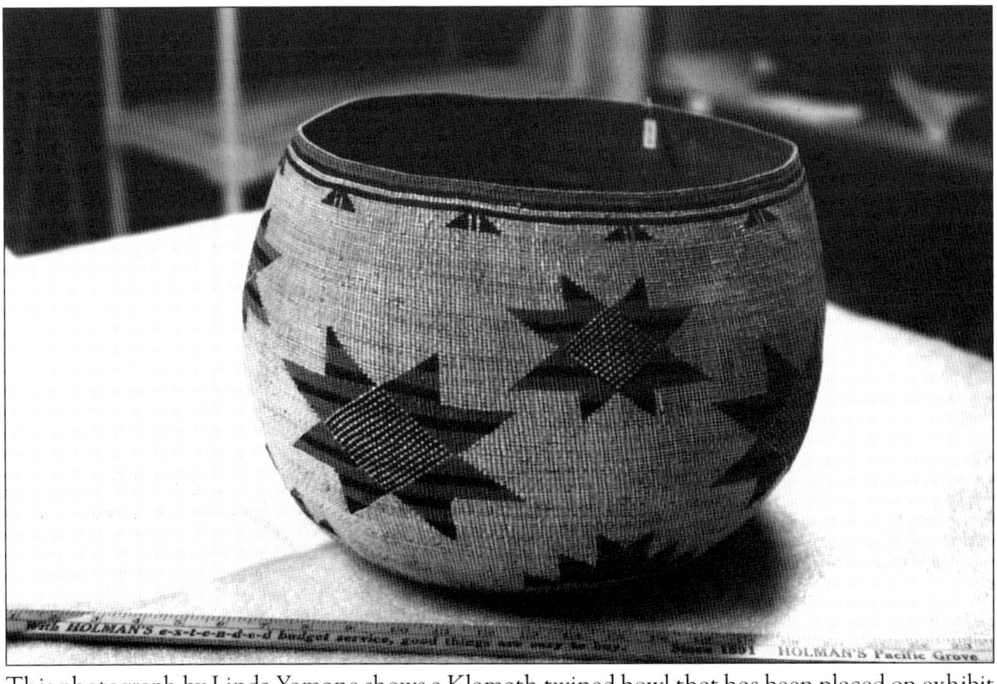

This photograph by Linda Yamane shows a Klamath twined bowl that has been placed on exhibit at the museum to illustrate the life of the Central Coast's indigenous peoples. Yamane continues to work with the museum on exhibits and can be viewed on a museum video as she describes the process of basket making. (Courtesy Pacific Grove Museum of Natural History.)

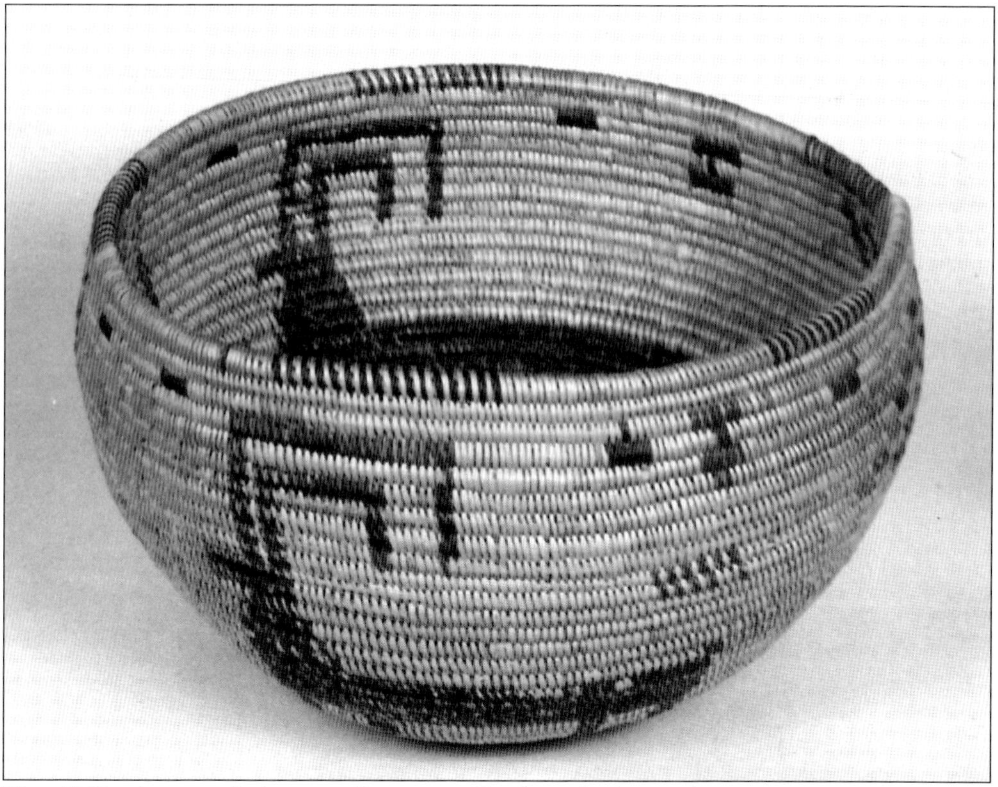

This 1960s photograph depicts a Panamint Shoshone basket with a Desert Big Horn design, as evidenced by the appearance of the animal inside the basket. The basket is made from several materials, including willow shoots, Joshua tree root, and bulrush root. Museum records indicate this object came from Death Valley in California. (Courtesy Pacific Grove Museum of Natural History.)

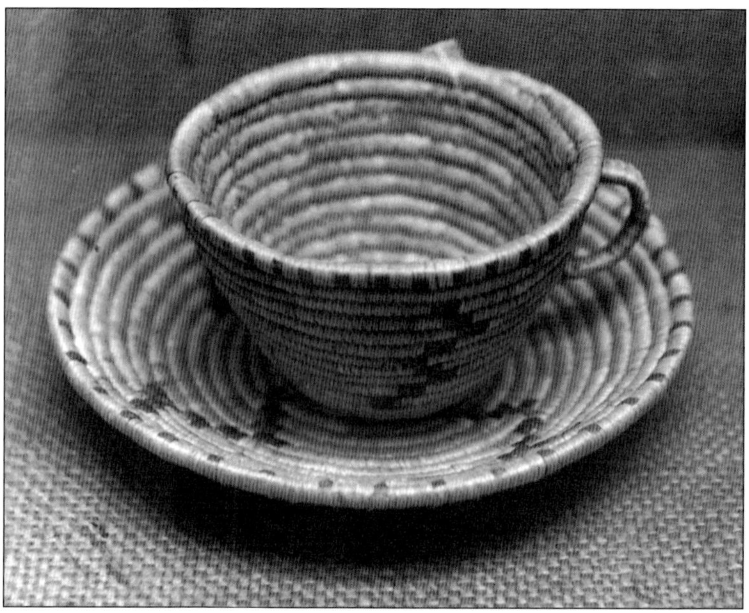

This item was placed on exhibit in August 1976 as part of the museum's basketry exhibit to illustrate the different forms of basket making found in the cultures of indigenous peoples throughout the Central Coast, the Southwestern states, and elsewhere. This image shows a cup and saucer made from basket weaving. (Courtesy Pacific Grove Museum of Natural History.)

This photograph by Linda Yamane features a *tubatulabal*, a coiled bowl, used in a museum exhibit in the 1990s. The item, as well as a host of others, were part of a California Indian basket exhibit, on display at the time. The museum will often contract items to exhibit that are not accessioned into the permanent collection but loaned to the museum for use. (Courtesy Pacific Grove Museum of Natural History.)

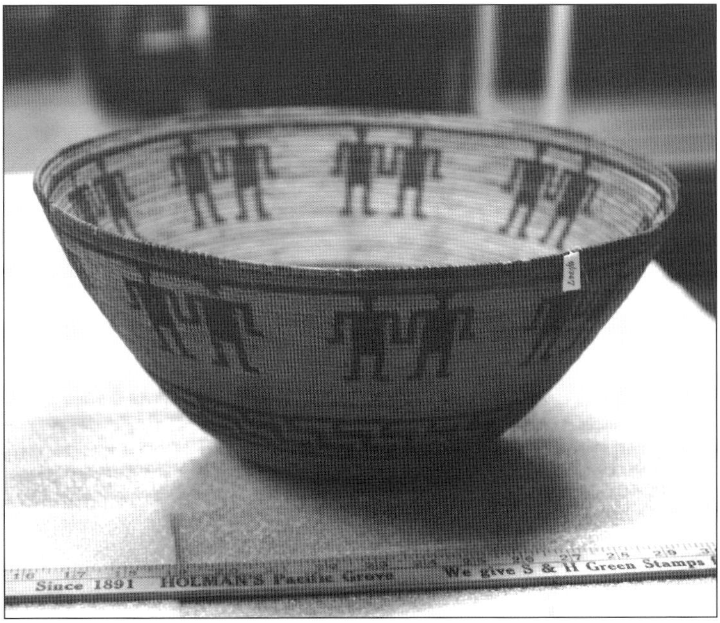

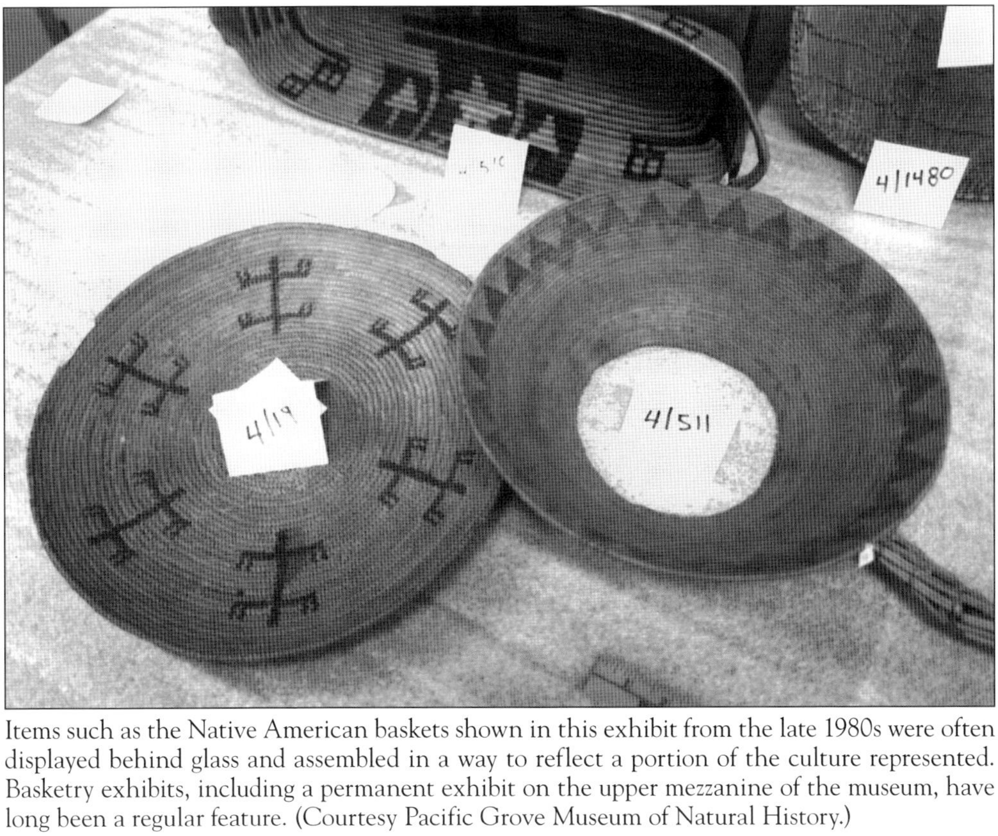

Items such as the Native American baskets shown in this exhibit from the late 1980s were often displayed behind glass and assembled in a way to reflect a portion of the culture represented. Basketry exhibits, including a permanent exhibit on the upper mezzanine of the museum, have long been a regular feature. (Courtesy Pacific Grove Museum of Natural History.)

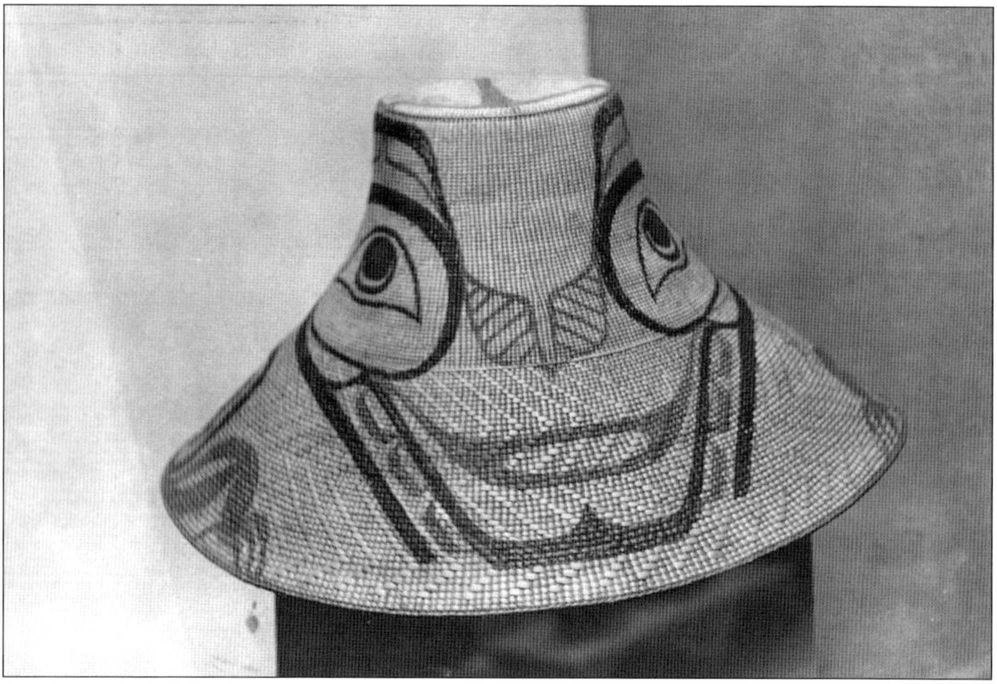

Exhibits included the Sun Art exhibit in 1976, as seen in this photograph taken at the time of the display. This item and others were placed behind glass and offered for viewing by the public. At the time, when the City of Pacific Grove operated the natural history museum, entry fees were waived, making the museum's admission free. (Courtesy Pacific Grove Museum of Natural History.)

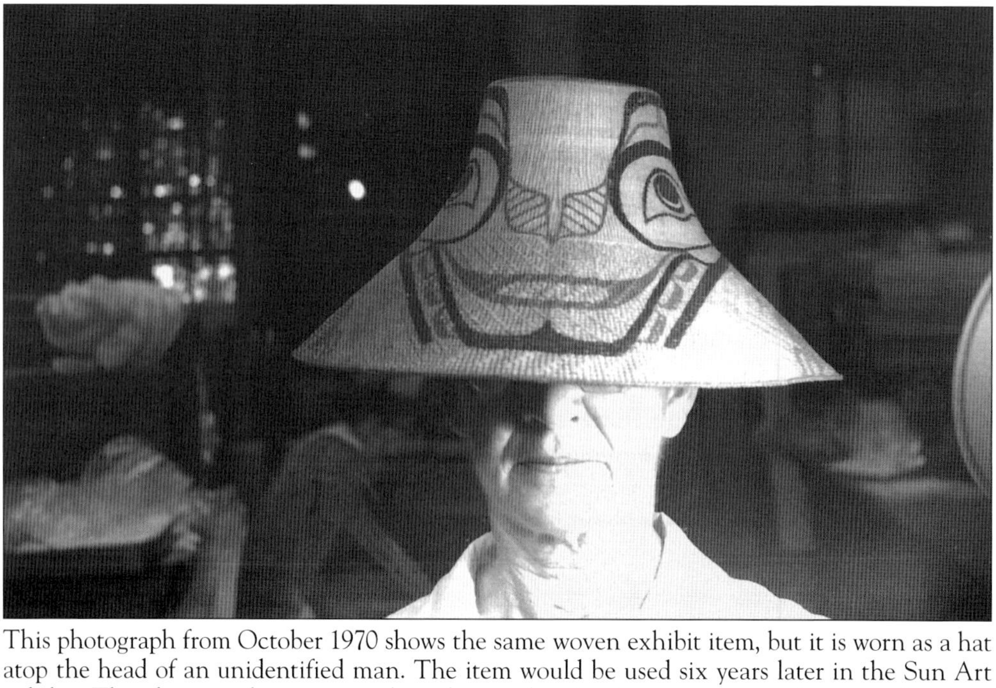

This photograph from October 1970 shows the same woven exhibit item, but it is worn as a hat atop the head of an unidentified man. The item would be used six years later in the Sun Art exhibit. The photograph appears to have been taken in the museum's work room. (Courtesy Pacific Grove Museum of Natural History.)

Pictured is a Haida "Copper," or dead plat totem, perhaps originally part of a ship's boiler, from the museum's basement collection. This item was collected around 1903 by Dr. Harold Heath, a prominent part of the museum, serving in the role of director from 1916 to 1947. Heath died in 1951. (Courtesy Pacific Grove Museum of Natural History.)

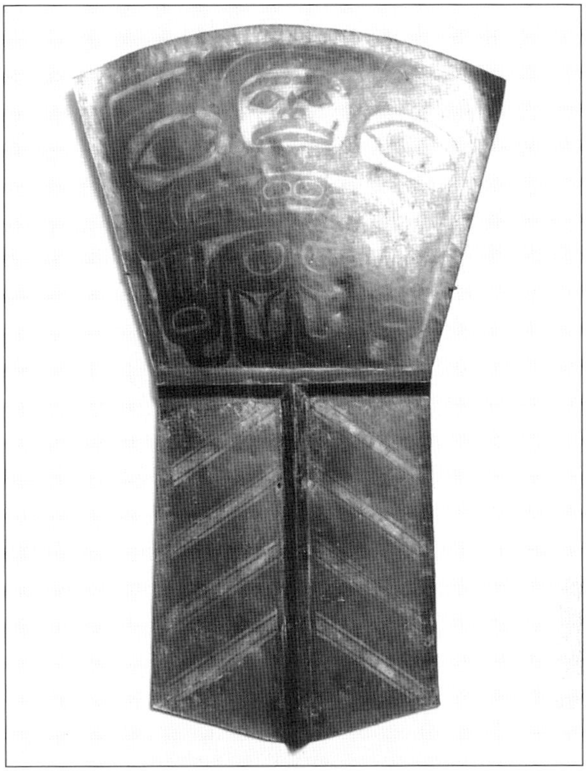

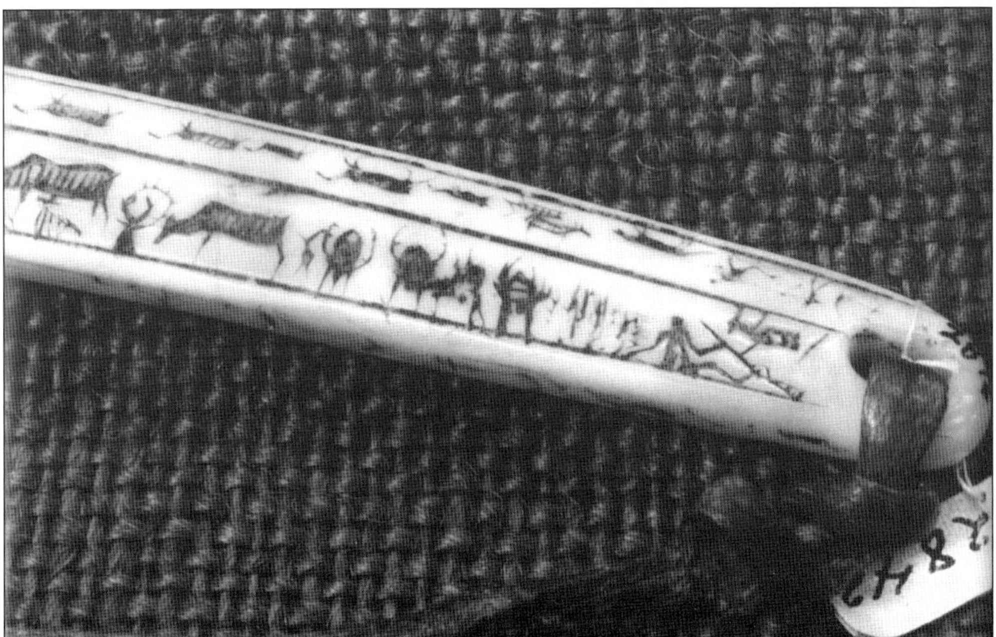

This photograph from 1937 shows an etched walrus tusk with hunting scenes, likely Inupiaq, carved into it. The tusk was collected by C.B. Phillips and brought to the museum for safe keeping. While many items in the museum collection are not displayed, they are kept in the basement under careful care. (Courtesy Pacific Grove Museum of Natural History.)

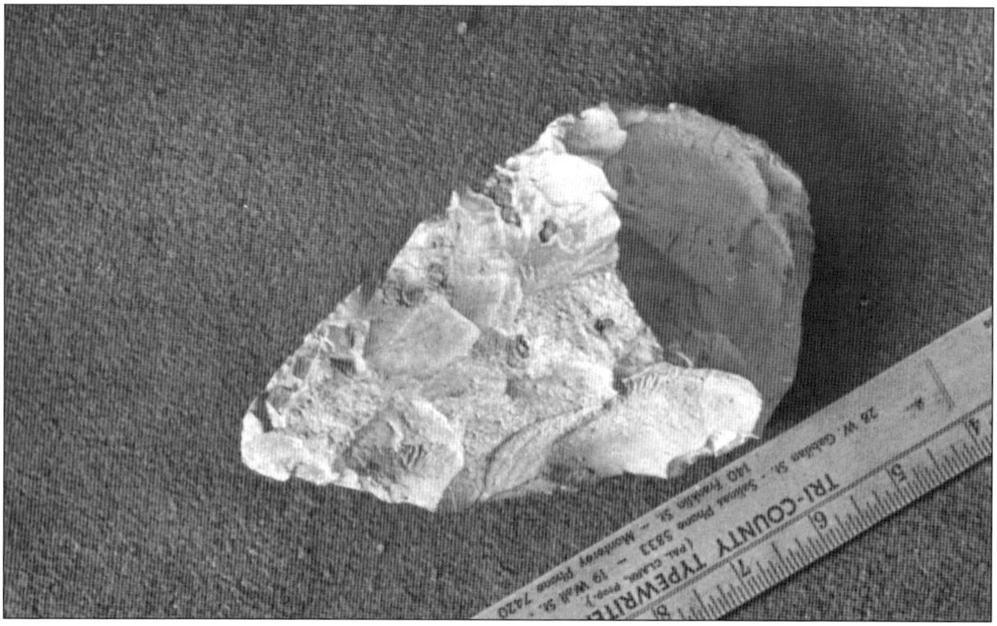

This arrowhead from Coyote Lake, California, was collected by Winifred Stewart in the 1970s. Coyote Lake can be found inland from the Central Coast near the city of Morgan Hill. Exhibit items such as this one were often photographed with a ruler to easily reference the size of the item. (Courtesy Pacific Grove Museum of Natural History.)

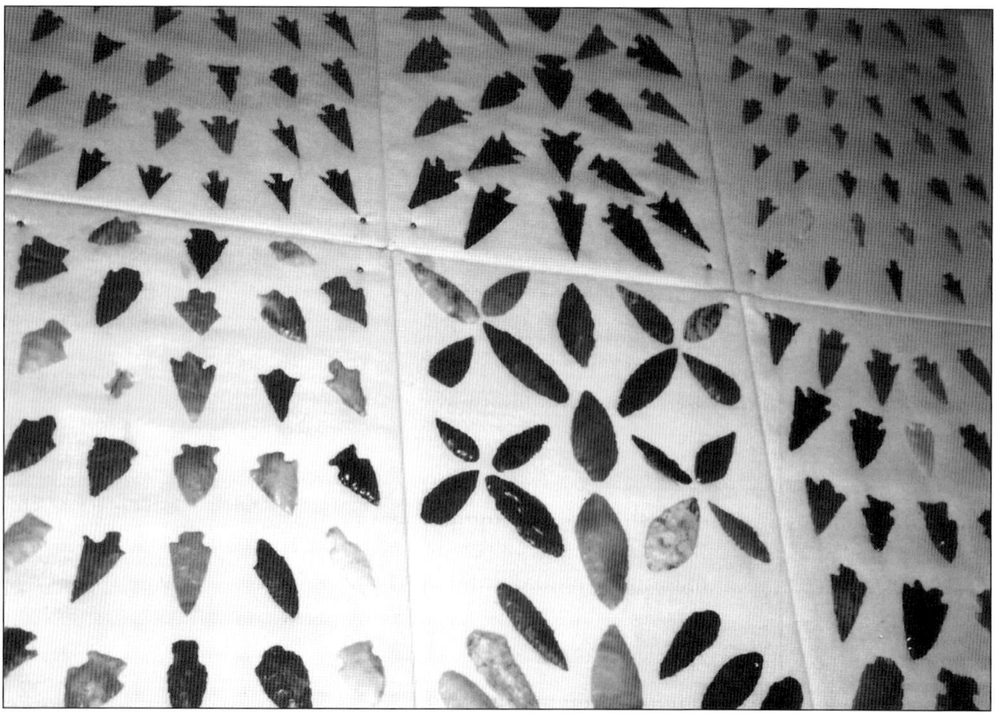

The museum has a collection of thousands of Native American lithic tools, both local and from throughout the western states. These items are gathered and stored for safe keeping in containers housed in the museum's large basement. (Courtesy Pacific Grove Museum of Natural History.)

This photograph from 1969 shows a relief map of the Monterey Bay, detailing the contours of the bay's deepest sections. The framed map has been on display at the museum for decades. A second, larger map is also part of the museum's regular exhibit in the Heritage Gallery. (Courtesy Pacific Grove Museum of Natural History.)

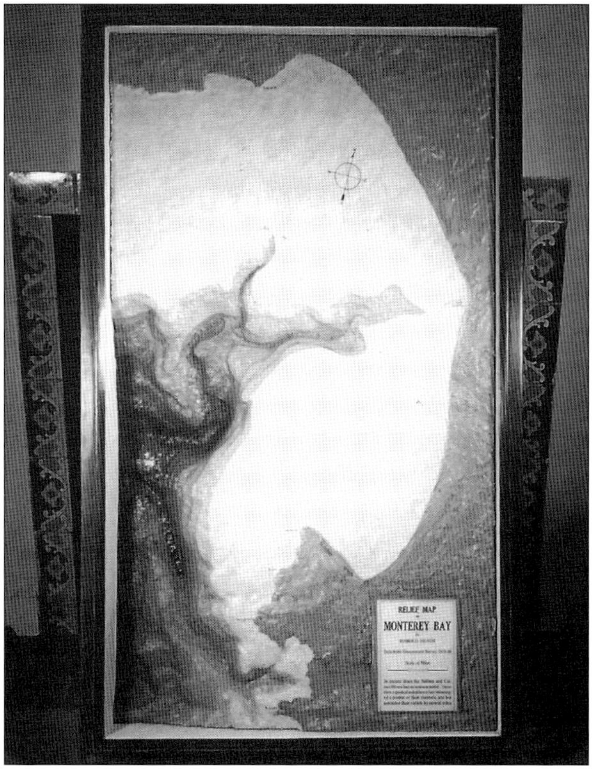

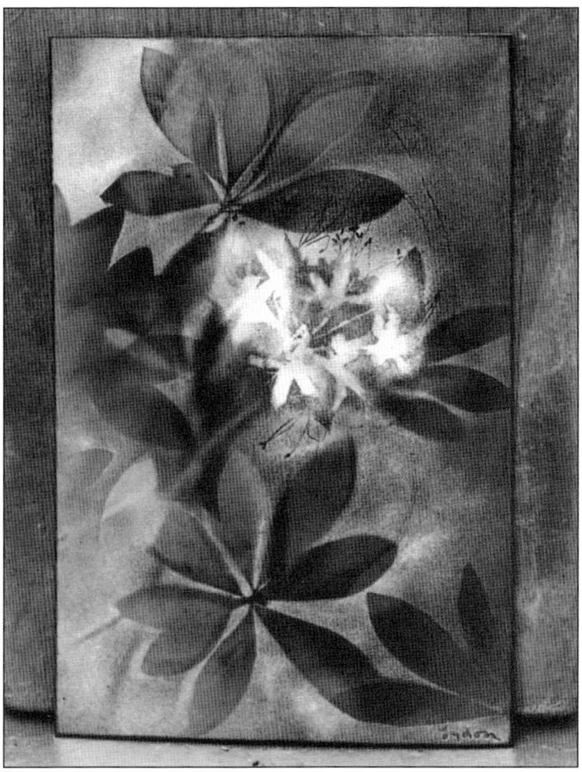

This enamel on copper panel art by Kay Lydo and photographed in the late 1960s features a *Rhododendron occidentale* (western azalea). The piece of art, which measures eight inches wide by almost 12.5 inches tall, is fastened in a raised position on a piece of gray-toned wood. (Courtesy Pacific Grove Museum of Natural History.)

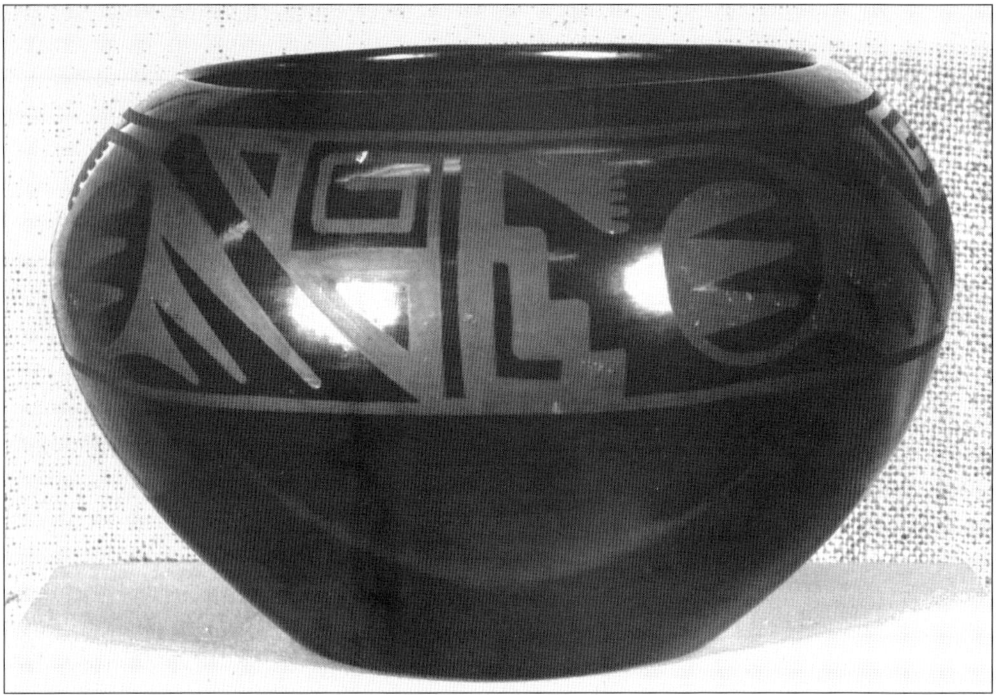

This photograph from 1974 is of a San Ildefonso Pottery item measuring six and a half inches tall. A signature is incised in the piece of pottery that reads, "Marie and Julian." The pottery is also referred to in the curator's notes as the Maria Martinez pot. The pot is pictured behind a weaved fabric, possibly as it appeared on exhibit. (Courtesy Pacific Grove Museum of Natural History.)

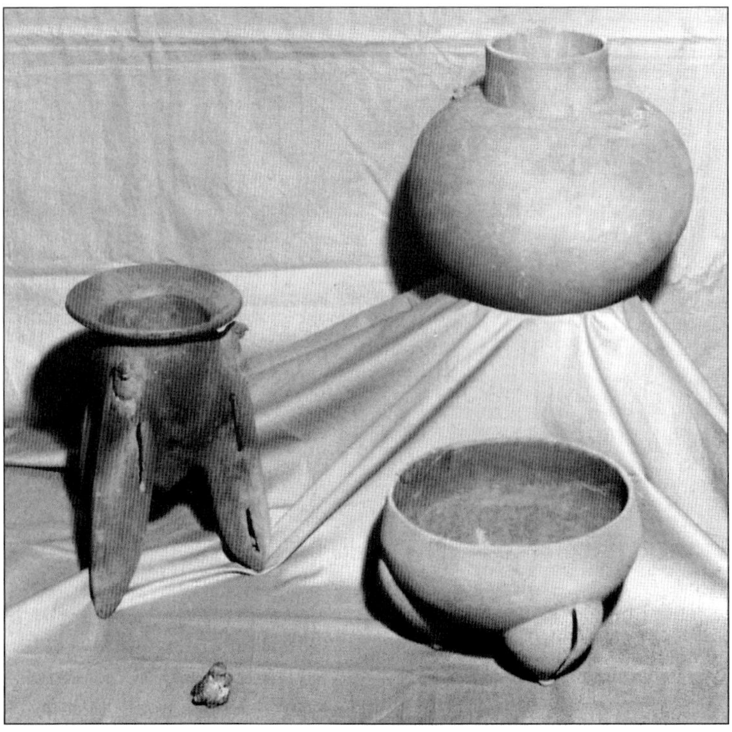

This photograph from the museum's collection was taken in 1956 and shows a set of four pottery items collected from the Chiriqui Province in Panama. The province borders Costa Rica and the vast Pacific Ocean. The items are shown as they appeared either on exhibit or displayed solely for photographic purposes. Often, museum items are photographed with white backdrops to avoid distractions in the image. (Courtesy Pacific Grove Museum of Natural History.)

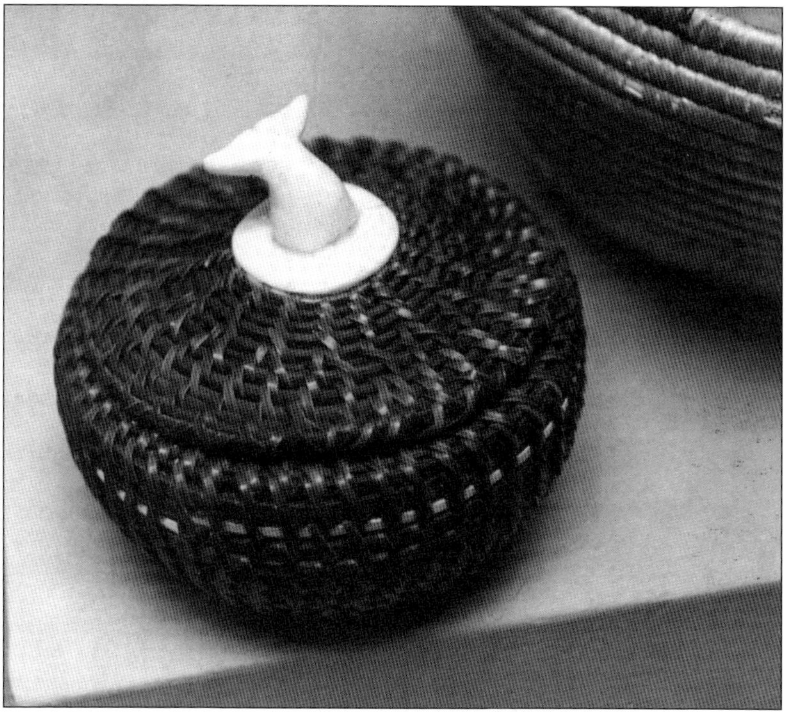

This photograph from the late 1970s is an ornate example of Eskimo basketry, according to museum records. Over the decades, it has been a regular exhibit item. The item, which is smaller than it appears in this image, is pictured next to another basket of a larger size. (Courtesy Pacific Grove Museum of Natural History.)

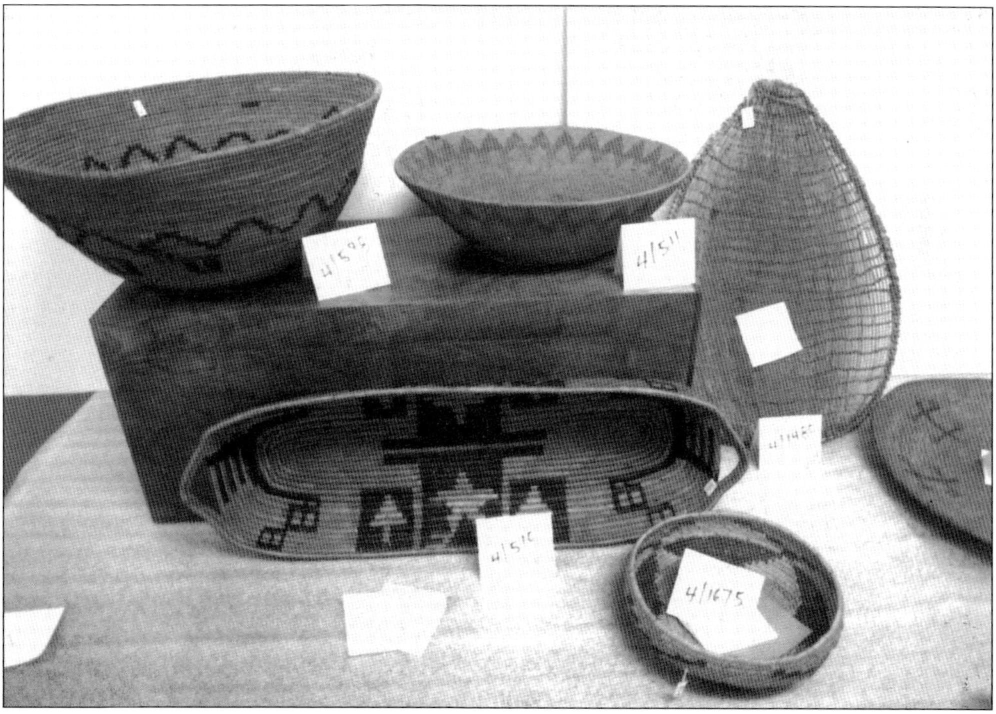

Pictured here is another example of Native American baskets as they appeared during a late 1980s exhibit at the museum. The items are arranged carefully for display to allow optimal viewing from the public. Openings for exhibits of this kind can, at times, draw large crowds to the natural history museum. (Courtesy Pacific Grove Museum of Natural History.)

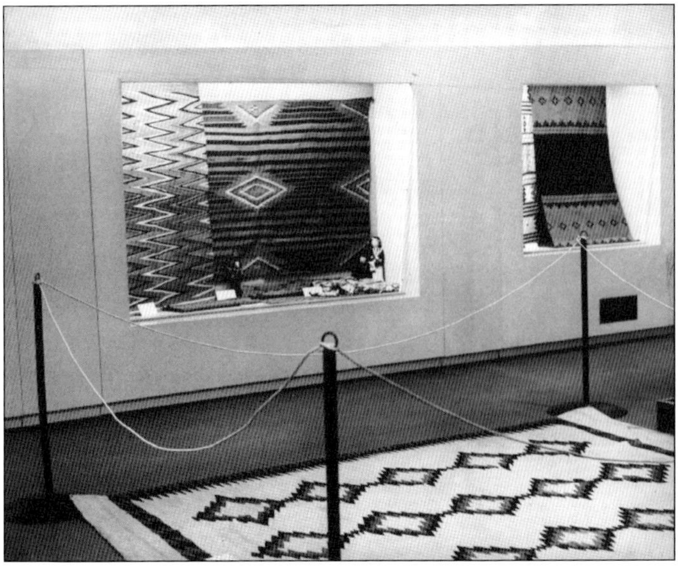

This Sun Art exhibit photograph from 1979 shows a collection of Navajo weavings. The exhibit was shown in the Room of Man, an area no longer named such in the museum. Today, the Room of Man is known as the special exhibits gallery. Some of the items, placed on the floor, are cordoned off in the image. (Courtesy Pacific Grove Museum of Natural History.)

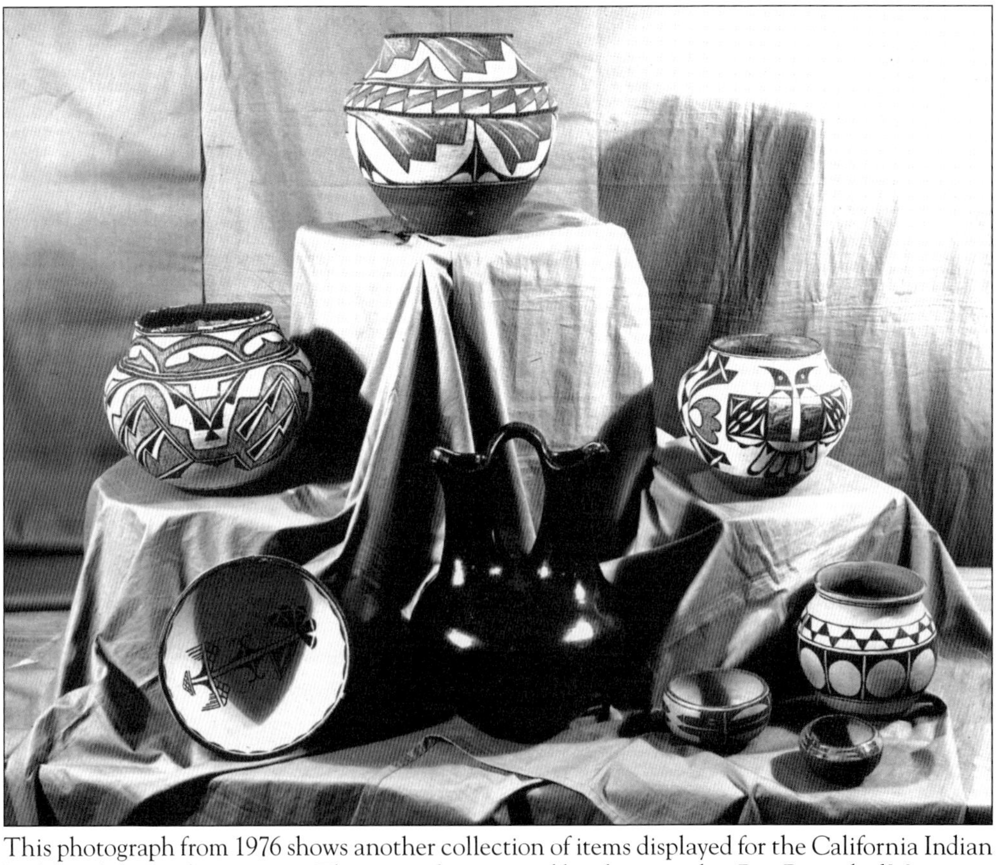

This photograph from 1976 shows another collection of items displayed for the California Indian basket exhibit in the museum. The image for captured by photographer Rey Ruppel of Monterey. Ruppel took several photographs for the Pacific Grove Museum during this period. Several different types of pottery, from small to large, are visible. (Courtesy Pacific Grove Museum of Natural History.)

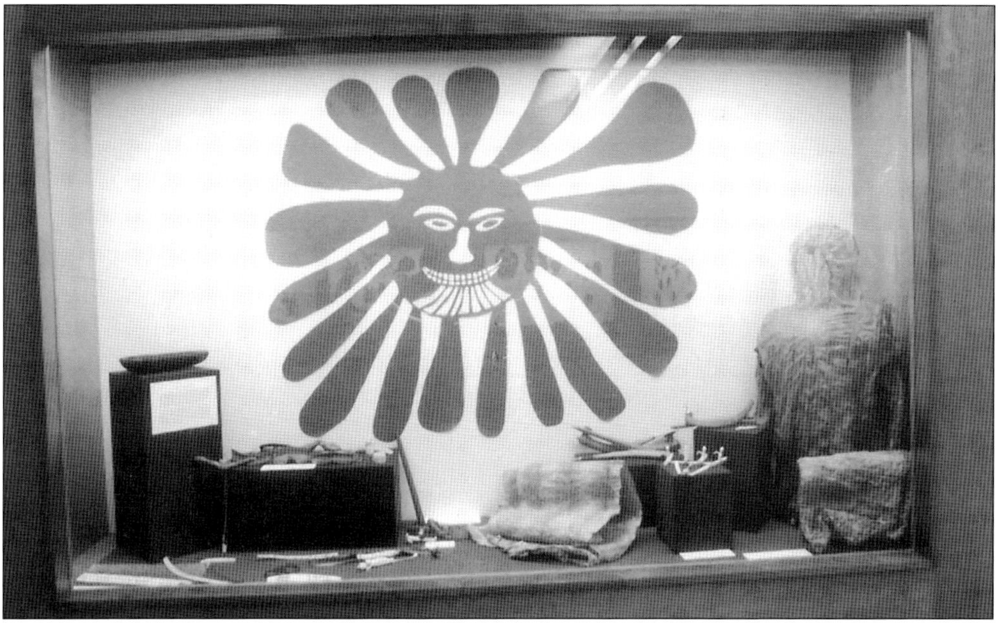

Pictured is another Native American display photographed in 1991, featuring many intriguing exhibit items. Placed behind protective glass, this exhibit featured a painted sun overlooking the collection. To portray the items to maximum effect, many were placed on exhibit boxes and raised above other items. (Courtesy Pacific Grove Museum of Natural History.)

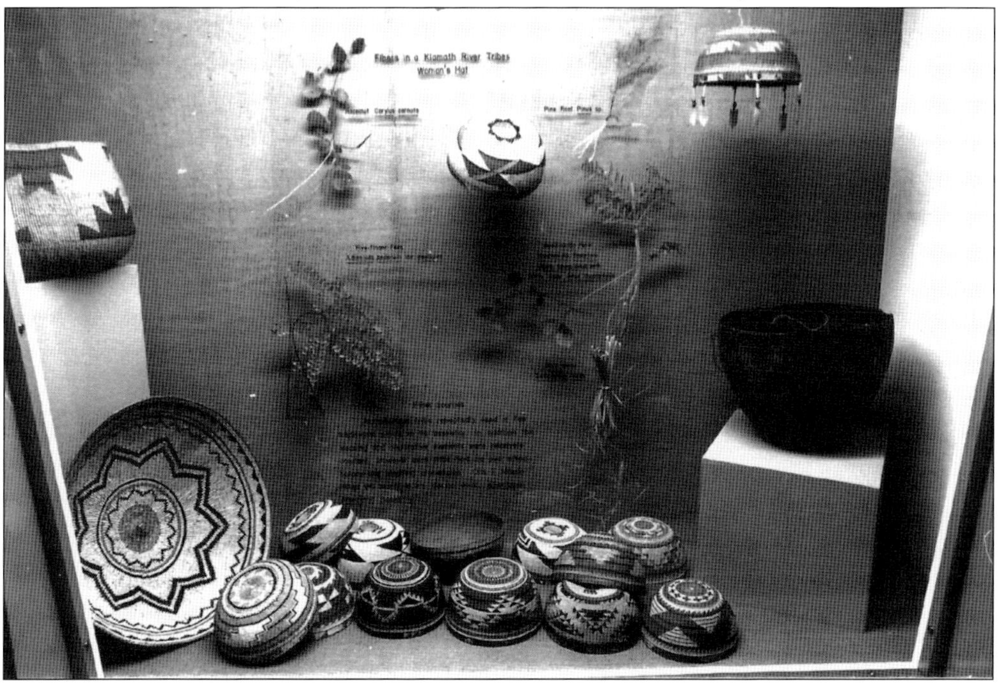

This image from the 1960s shows an exhibit dedicated to the Klamath River tribes. Pictured are examples of the plants whose fibers were used to make many of the basket exhibits also displayed in the museum exhibit. The text in this display explains how the fibers were used to create the baskets and hats. (Courtesy Pacific Grove Museum of Natural History.)

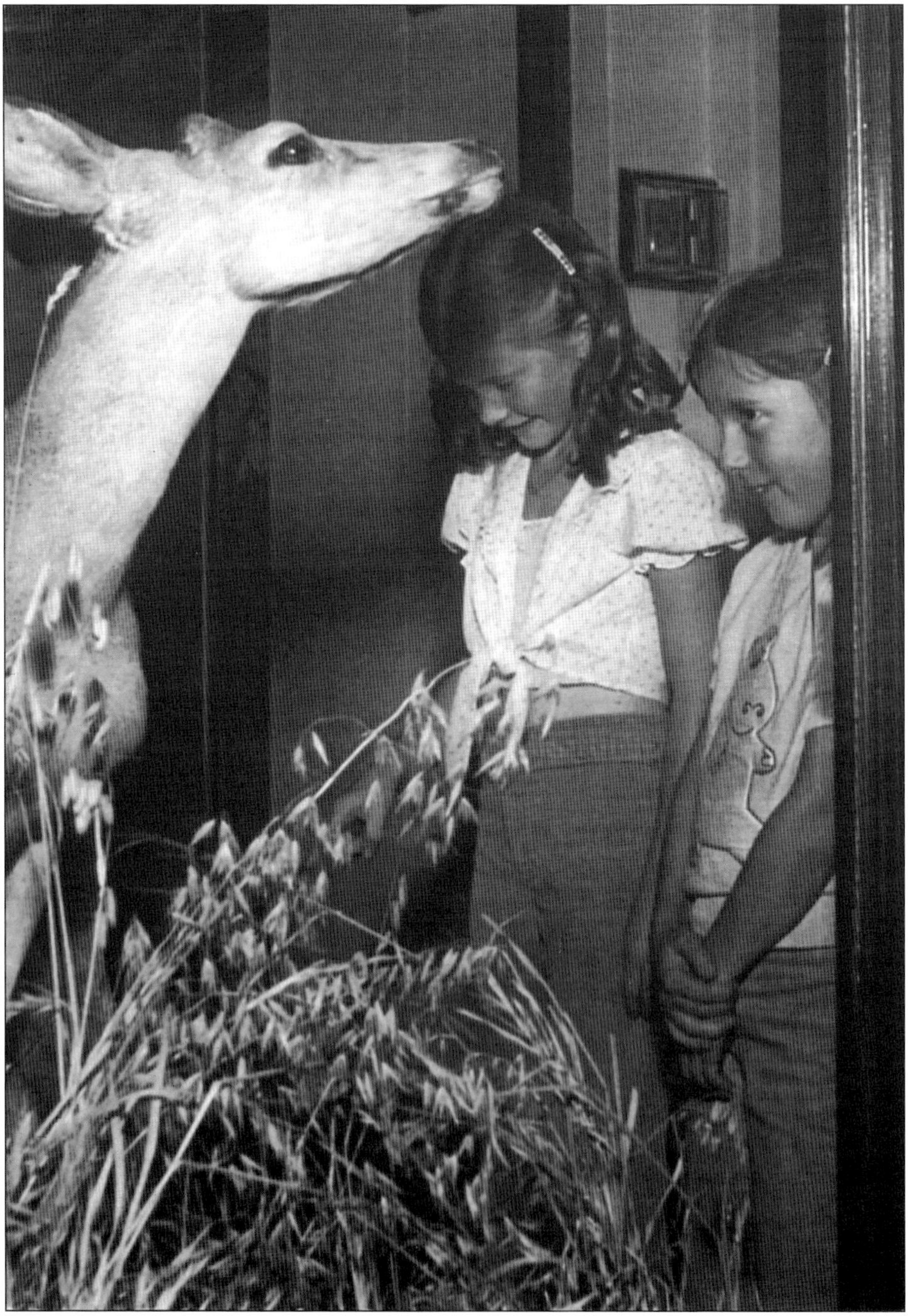

This photograph from the early 1970s shows two young girls looking at the deer case diorama in the main room of the original museum, which is today known as the Heritage Gallery. The case contains two male deer (spiked fawns) behind glass, part of the museum's Mammalia collection, with a backdrop painted by artist Albert DeRome. One deer is visible looking toward the sky, much to the delight of the two girls, while the second deer, which cannot be seen in this image, is fixated on the vegetation surrounded them on the floor. Besides the two deer, there are currently two life-mounted bears in the Heritage Gallery. (Courtesy Pacific Grove Museum of Natural History.)

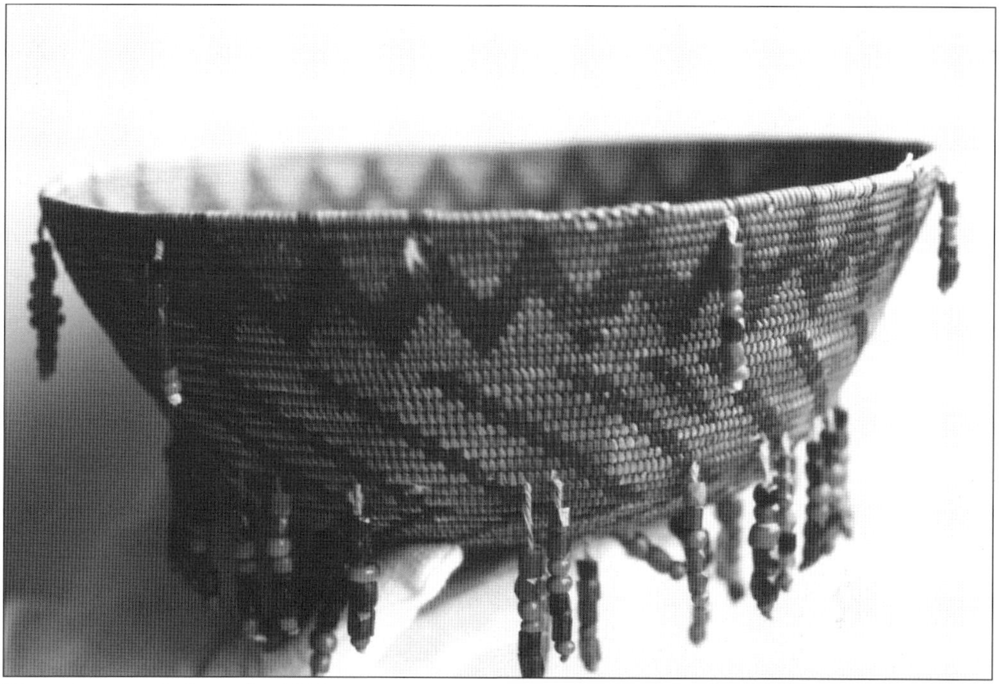

This photograph from the 1990s is of a Linda Yamane Salinan basket, elaborately made with color interspersed throughout the basket. Yamane's baskets have long been on exhibit at the museum and continue to be a part of the basket exhibit collection today. The photograph shows a gloved hand holding the basket. This prevents oils from the skin from contacting the object. (Courtesy Pacific Grove Museum of Natural History.)

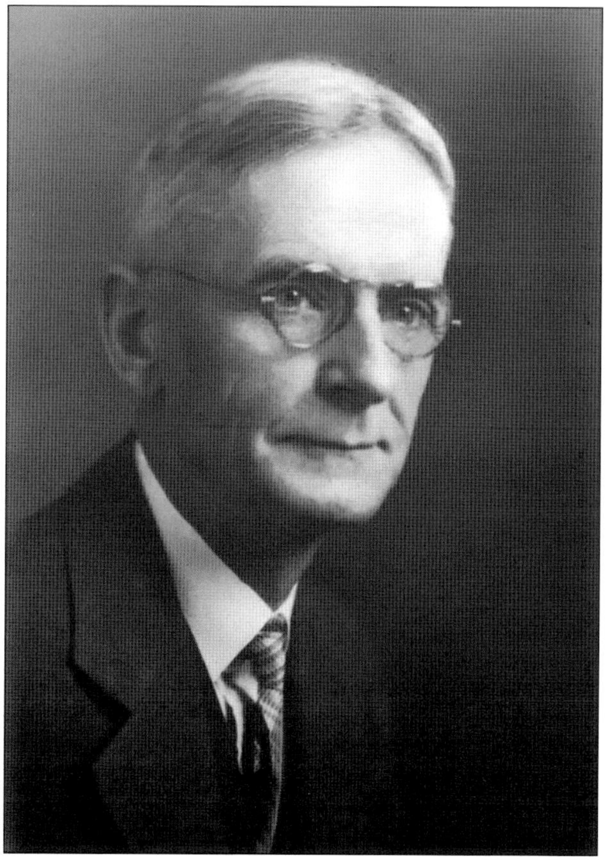

Pictured is Dr. Harold Heath in the late 1940s. Heath was well known for his many scientific expeditions to exotic locales, including Alaska, among others. Besides his work as a professor of zoology at Stanford, Heath also served on the museum's board of trustees from 1916 to 1917. He worked as the museum's scientific director from 1916 to 1947. (Courtesy Pacific Grove Museum of Natural History.)

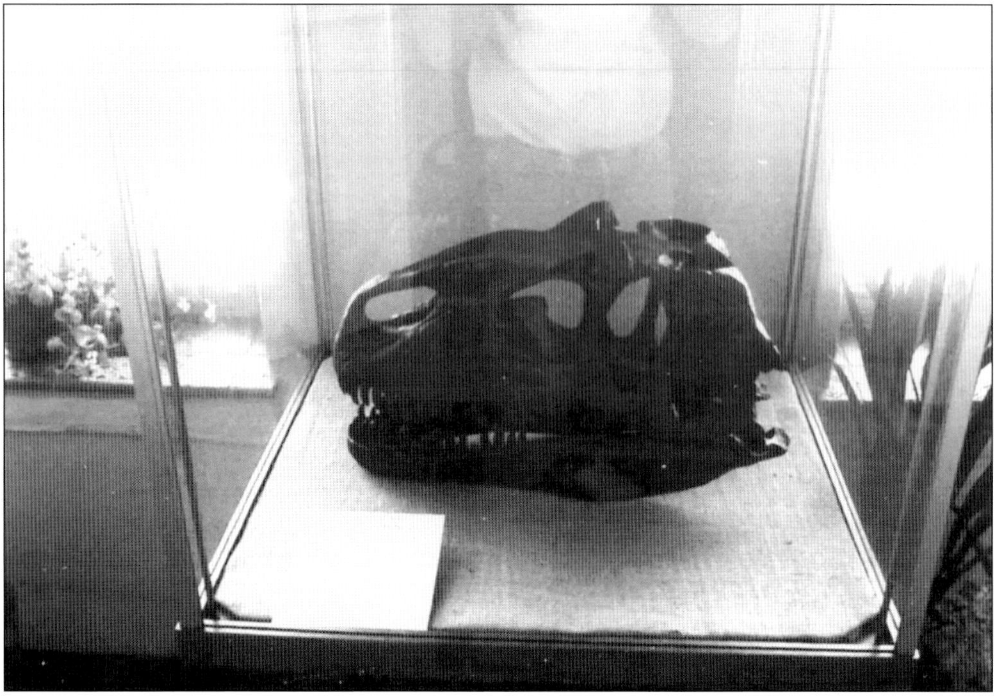

This photograph from 1986 shows a cast of an *Allosaurus fragilis* cranium. The cranium is black in color and was made from an original specimen found at the Utah Museum of Natural History. The original cranium was found in Utah and dates to the late Jurassic Era (150 million years ago). (Courtesy Pacific Grove Museum of Natural History.)

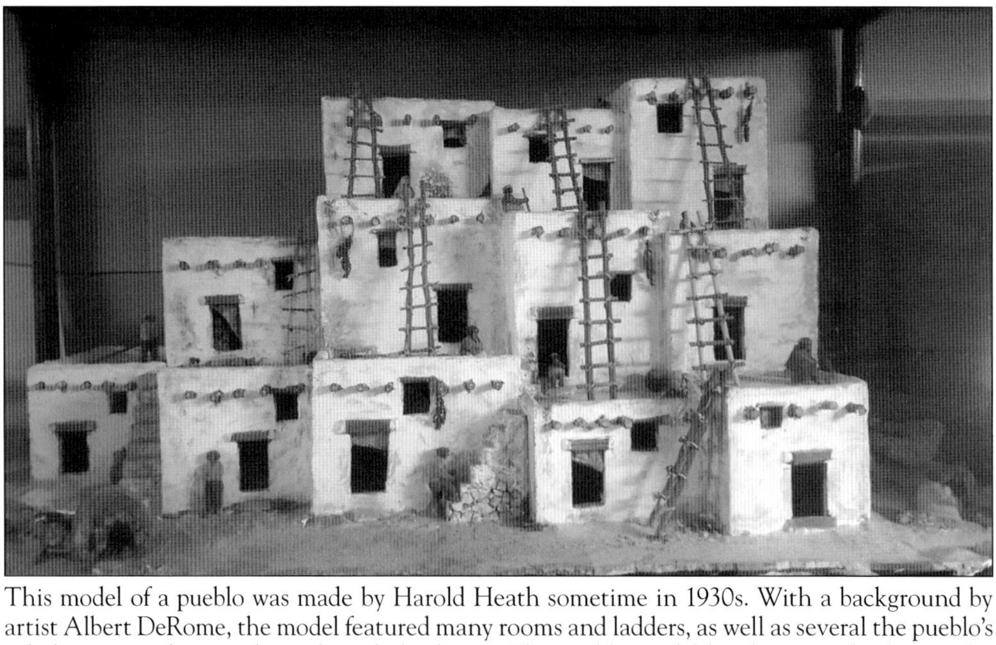

This model of a pueblo was made by Harold Heath sometime in 1930s. With a background by artist Albert DeRome, the model featured many rooms and ladders, as well as several the pueblo's inhabitants as they go about their daily chores. The pueblo model has been on display on the museum's mezzanine level and is currently stored as part of the museum's basement collection. It is kept in a wooden case with glass panels. (Courtesy Pacific Grove Museum of Natural History.)

Four

A Part of the Community

Local institutions, with a shared connection to the world of natural history, have partnered with the Pacific Grove Museum of Natural History on either events or as a place to hold lectures and workshops since the museum began. The museum has also been a close partner with the City of Pacific Grove and the local Pacific Grove Chamber of Commerce and remains so today. Both the museum and the chamber originally began in octagonal structures of similar design, though some historians believe the chamber's building was originally a beach house while the museum's original structure was a smoking patio for the El Carmelo Hotel. The early Sanborn maps, depicting various locations along Pacific Grove, including the hotel and Jewell Park before and after the construction of Central Avenue, seem to confirm this belief. Support also came from a number of individuals in the community, such as Julia Platt, an early conservationist who helped create protections for the shoreline around Pacific Grove. Platt also planted gardens at Lovers Point and drafted plans for walkways and plant improvements along the shoreline of the city. In 1931, she became the first female mayor of Pacific Grove. Her shoreline improvement drawings, meanwhile, eventually found a home at the museum. The city's first mayor, Oliver Trimmer, served for 20 years. He also functioned as a director on the museum's board from 1900 to 1903. The museum has played host to several nonprofit organizations over the years, including the Carmel Valley Gem and Mineral Society, the Monterey Chapter of the Audubon Society, and the Monterey Bay Chapter of the California Native Plant Society (CNPS). The CNPS formed in 1965, with the forming of the Monterey Bay chapter a year later. For more than 50 years, the organization has partnered with the museum for their annual wildflower show, which features plant samples gathered from numerous locations across the county and displayed for visitors to see. The fair has been advertised as the largest wildflower show in the western hemisphere.

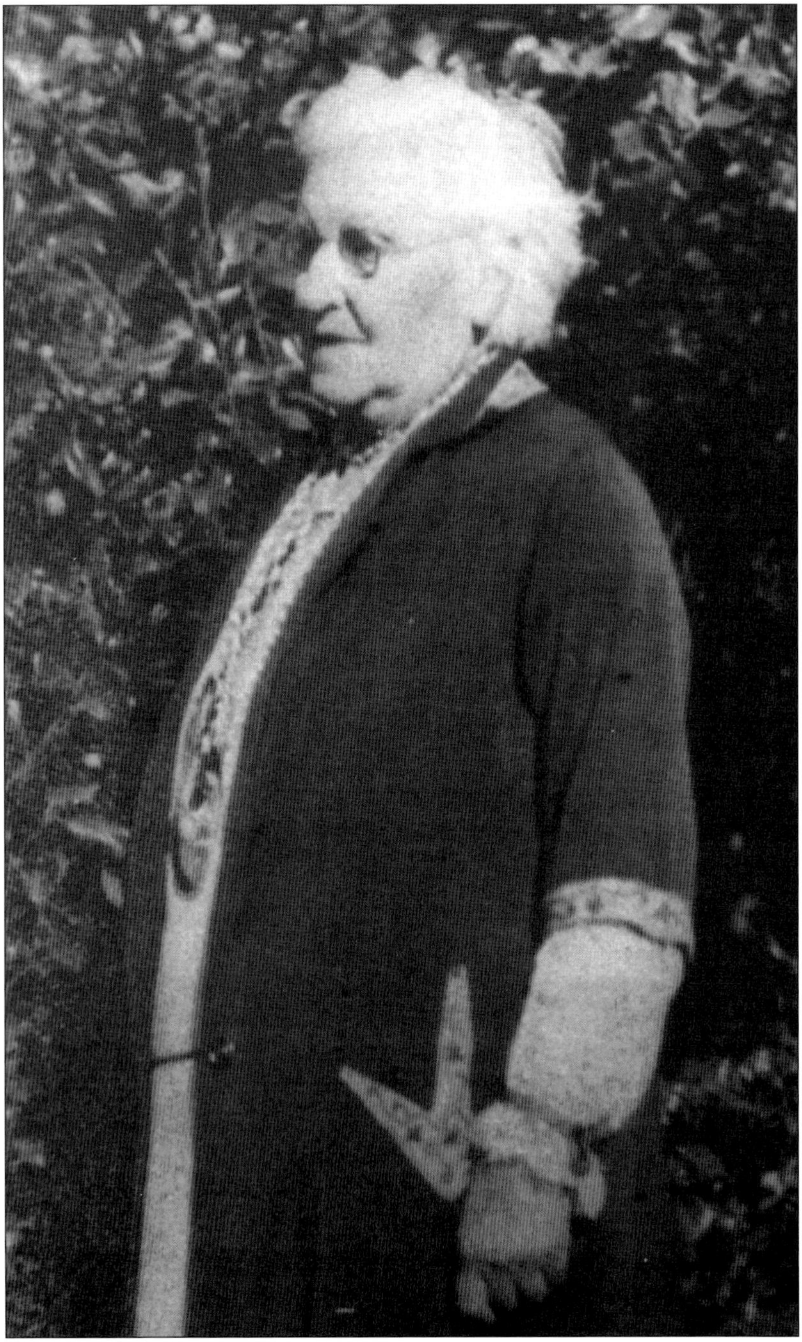

Julia Platt is pictured in the 1930s. At the age of 74, she became the first female mayor of Pacific Grove in 1931. A noted lover of Monterey Bay and the shoreline of Pacific Grove, Platt is credited with helping create the sanctuary shores around Pacific Grove. Many of her drawings, for natural spaces along the shore, can be found in the museum's collection. Platt shared a love for the Central Coast found in many of the residents of Pacific Grove at the time, including those who operated the natural history museum and collected items for future study that would be stored there. (Courtesy Pacific Grove Public Library.)

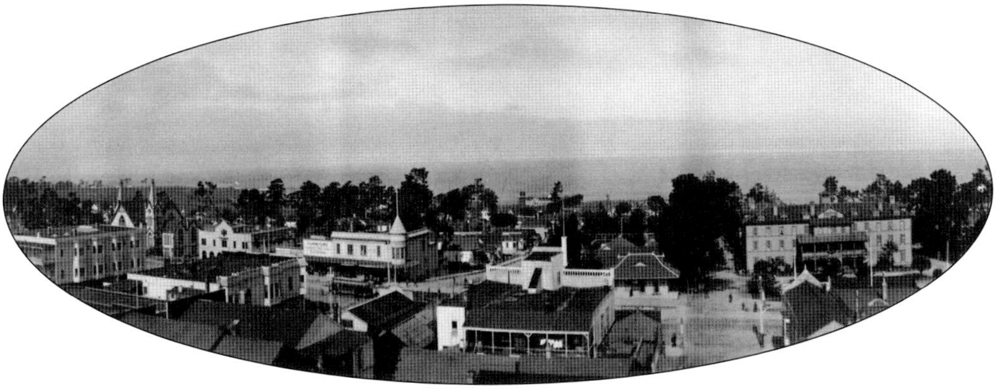

This oval panoramic of downtown Pacific Grove, taken in 1903 from the roof of Mammoth Stables on Laurel Avenue, offers a splendid view of the Pacific Grove community at a time when the museum continued to grow. Several buildings can be seen in the photograph, including the Hopkins Seaside Laboratory, the El Carmelo Hotel, the office of the Pacific Improvement Company, and more. The tennis court at Jewell Park is also visible. (Courtesy Pacific Grove Museum of Natural History.)

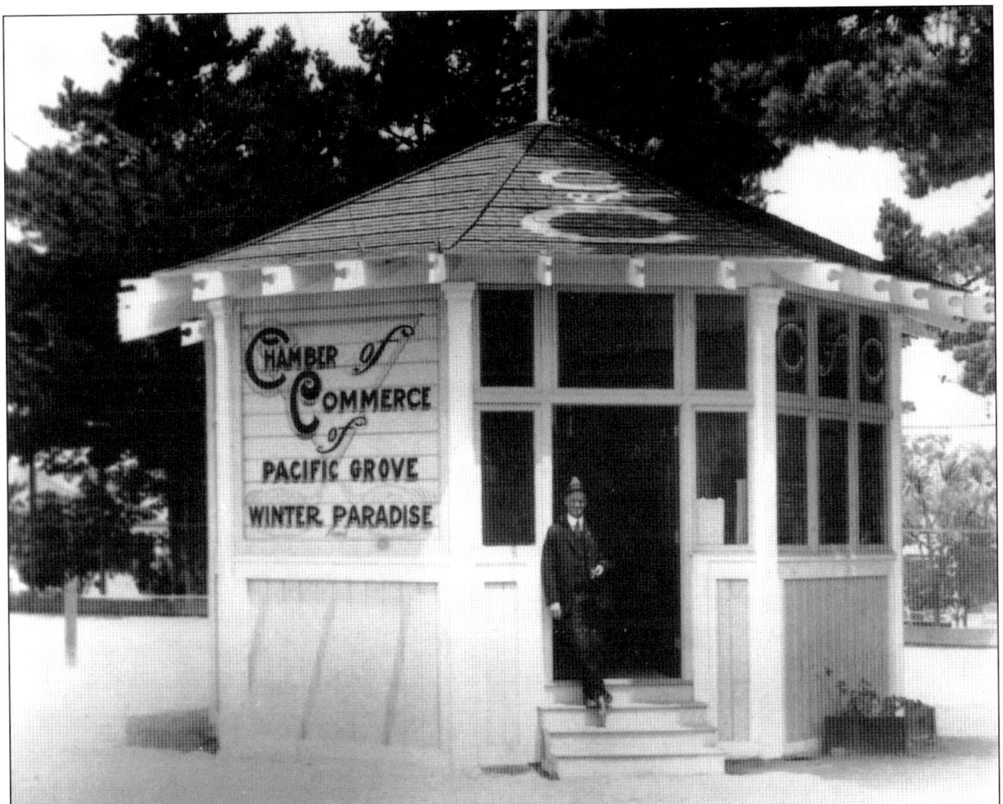

The museum was not the only octagonal building in Pacific Grove. As seen in this photograph from 1925, the Pacific Grove Chamber of Commerce got its start in just such a structure. It is seen here when the building was in Jewell Park near the tennis courts and across the street from the museum. Both remain neighbors to this day. The chamber began as the Board of Trade in the 1890s but later changed its name in 1915. (Courtesy Pacific Grove Museum of Natural History.)

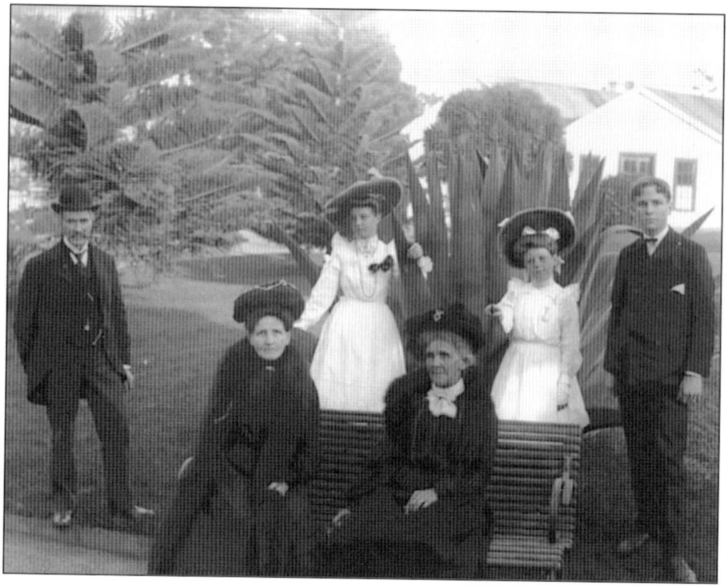

This 1906 photograph from the C.K. Tuttle collection housed at the museum features a group of six people in the garden of the El Carmelo Hotel, now the site of the Holman building. Pictured from left to right are Dr. O.S. Trimmer, Pacific Grove's first mayor and museum board member; Gladys Tuttle; Rhoda Trimmer; Winnifred Tuttle; and Floyd Tuttle. (Courtesy Pacific Grove Museum of Natural History.)

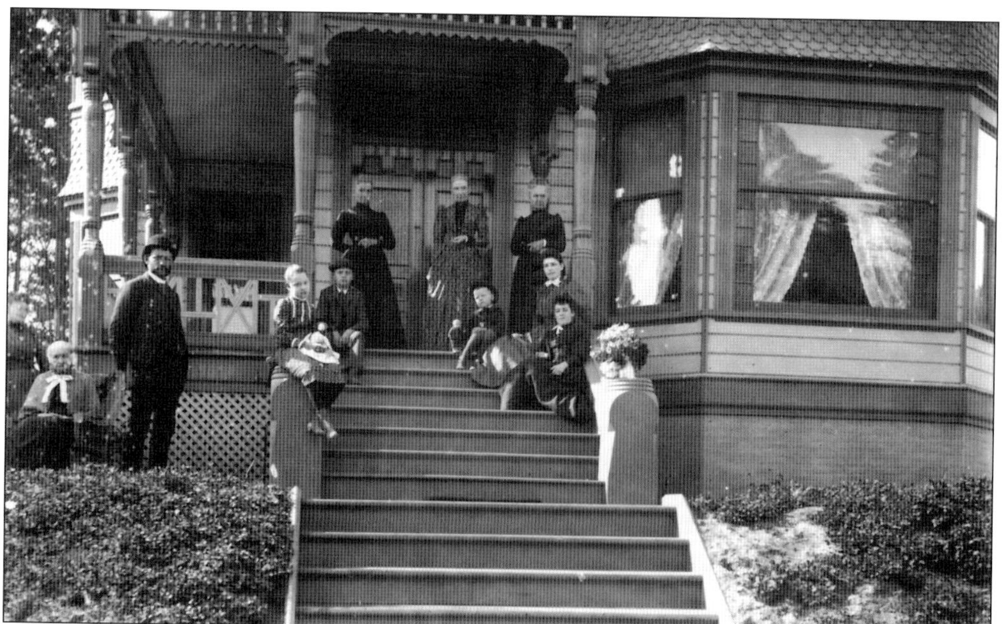

Pictured in 1895 are members of the Holman family gathered around their home on Lighthouse Avenue. The Holman family were called "pioneering" merchants in the Pacific Grove community. By 1924, their department store, located one block from the museum, would become the biggest such store between Los Angeles and San Francisco, bringing visitors to the site from miles away. Highway 68, also known as Holman Highway, was built in part to help customers to the store. (Courtesy Pacific Grove Museum of Natural History.)

This poster from 1979 advertises the annual Wildflower Show held at the Pacific Grove Museum of Natural History. For more than 50 years, the museum has partnered on the show with the California Native Plant Society's Monterey Bay chapter. The group still holds meetings at the museum every other month. (Courtesy Pacific Grove Museum of Natural History.)

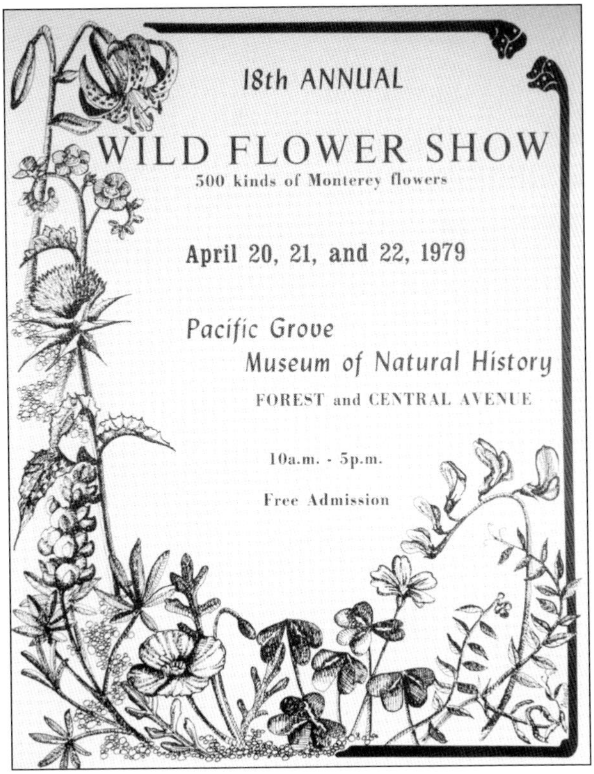

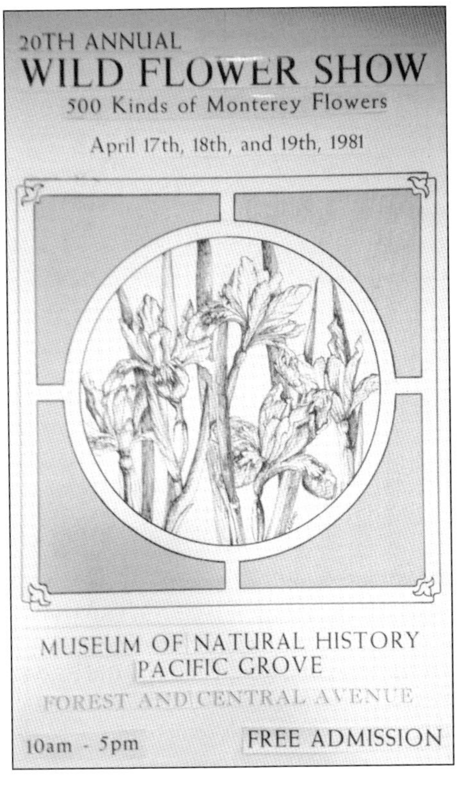

This is a mockup poster for the 1981 wildflower show at the museum. Organizers with the California Native Plant Society's Monterey Bay chapter scour the county for months prior to each show, collecting native plants to display during the three-day event. The 1981 event boasted more than 500 different plant samples. (Courtesy Pacific Grove Museum of Natural History.)

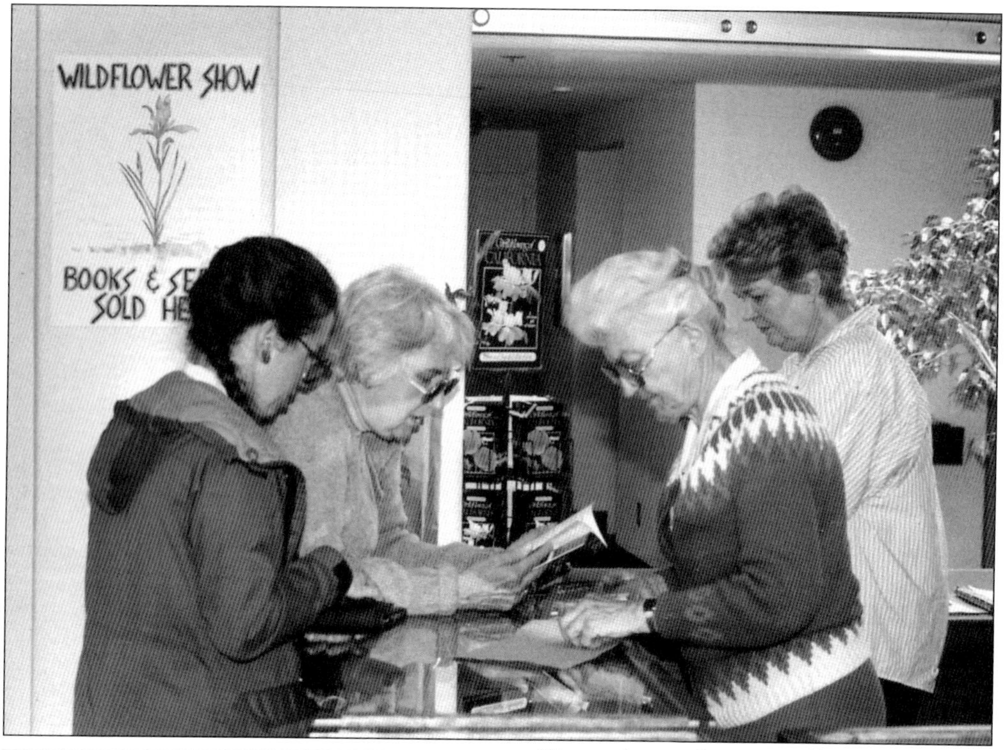

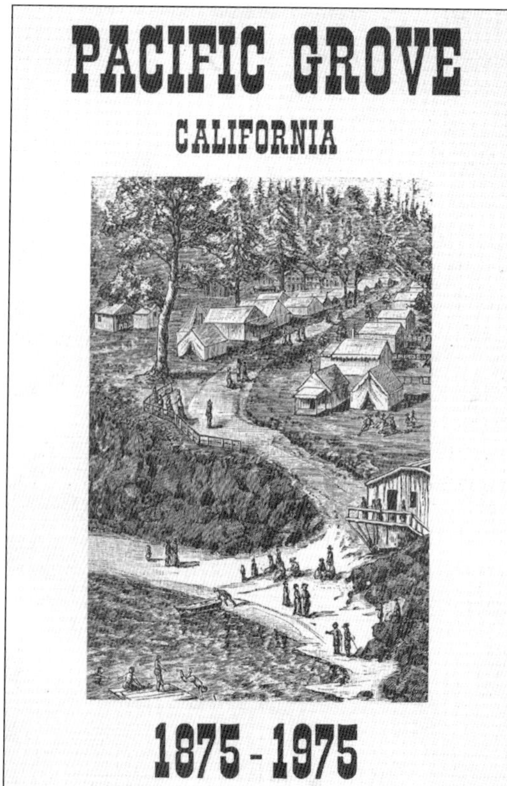

Pictured are volunteers selling native plant books during a later wildflower show held in the 1990s. One of the draws for this large annual event, held in the spring each year on the third weekend of April, is the books on native plants and gardening sold during the event. Today, the wildflower show is held in the bird gallery and offers more than 600 species and varieties of wildflowers. (Courtesy Pacific Grove Museum of Natural History.)

Pictured is the cover for a special program commemorating the 100-year anniversary of the city of Pacific Grove, from 1875 to 1975. The program featured "A Glimpse of the Past that Shaped the City's Present" and was produced by the Pacific Grove Centennial committee in June 1975 and featured the museum and other institutions that shaped the community. Pacific Grove mayor Robert A. Quinn designated the year 1975 as a centennial year for the city and made the week of May 25 through June 1 in 1975 Chautauqua Week. (Courtesy Pacific Grove Museum of Natural History.)

Five

GROWTH

The museum continued to grow after moving to its current location between Central Avenue and Lighthouse Avenue. Its collection grew significantly, as well. New additions included the apartments built in 1938 above the original entrance to the museum itself. The first to live there was the museum's first "official" curator, Teresa Lloyd, and her daughter Myfanwy. During that time, the curator was akin to the director of the museum. While the museum remained unchanged for decades after the construction of the penthouse apartments, today used as office space, the museum would receive valuable recognition over those years, including being named in 1959 as the "best museum of its size" in the United States by the National Association of Museums. The museum was also accredited by the American Association of Museums in 1972, the first such institution in Monterey County to do so. In 1980, plans were made to expand the building to its current size. These plans, first drafted in 1965, included a joint lecture and special exhibits space, and expanded exhibit space elsewhere in the museum. Other additions included handicap access and bathrooms. While the museum Association at the time raised $90,000 for the project, the additions were estimated to cost $329,000, much of which was paid for by grants and fundraising efforts. City council members, who originally denied the proposal in the 1960s, this time approved the construction request. The arrival of Sandy the Whale and the museum's subsequent efforts to purchase her helped raise museum awareness throughout the Monterey Peninsula and led to more than $600,000 in grants and donations, just enough for the new additions and the purchase of Sandy. The museum store, a gift shop for museum visitors, also opened in the 1980s. A year after it began in 1987, there were more than 43 volunteers tasked with keeping it open Tuesday through Sunday. Today, the museum features an elevator, multiple office spaces, a research library, an education room, the outdoor Native Plant Garden, and more.

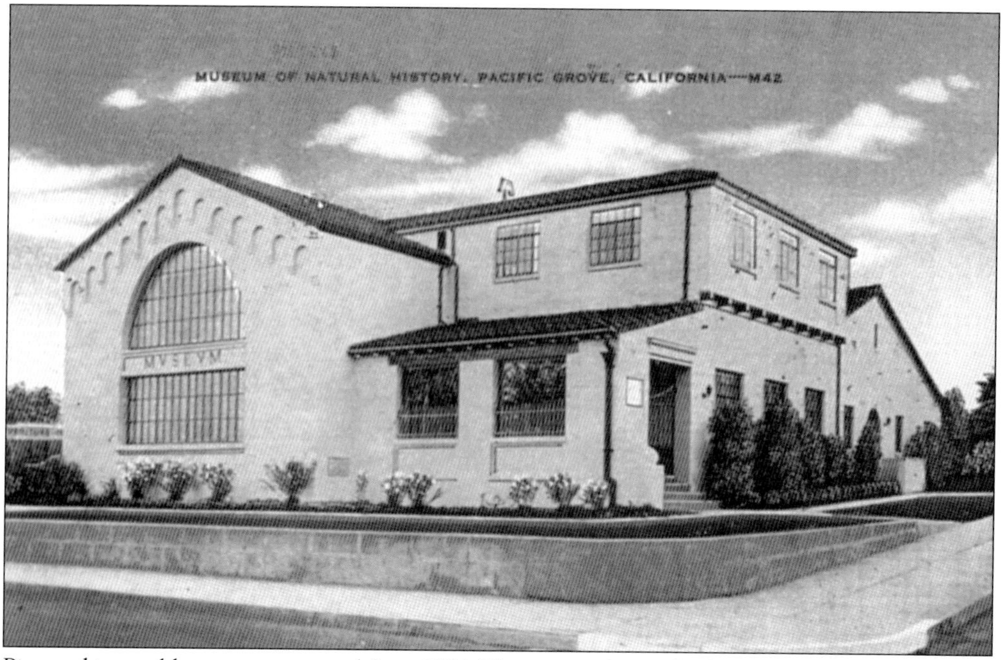

Pictured is an old museum postcard from 1934. The image shows the front of the museum, as well as the apartments built on the upper floor. The apartments served as the home for the curator and included a full bathroom and kitchen. The apartment is now used as offices for museum staff. (Courtesy Pacific Grove Museum of Natural History.)

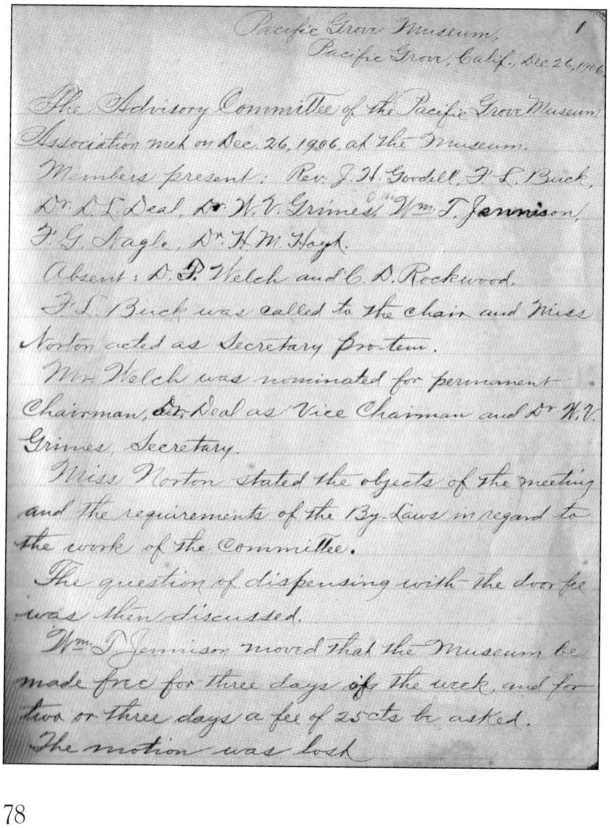

Pacific Grove Museum, 1
Pacific Grove, Calif., Dec. 26, 1906

The Advisory Committee of the Pacific Grove Museum Association met on Dec. 26, 1906, at the Museum.
Members present: Rev. J. H. Goodell, F. L. Buck, Dr. L. L. Deal, Dr. W. V. Grimes, Wm. T. Jennison, F. G. Nagle, Dr. H. M. Hoyt.
Absent: D. P. Welch and C. D. Rockwood.
F. L. Buck was called to the chair and Miss Norton acted as Secretary pro-tem.
Mr. Welch was nominated for permanent Chairman, Dr. Deal as Vice Chairman and Dr. W. V. Grimes, Secretary.
Miss Norton stated the objects of the meeting and the requirements of the By-Laws in regard to the work of the Committee.
The question of dispensing with the door fee was then discussed.
Wm. T. Jennison moved that the Museum be made free for three days of the week, and for two or three days a fee of 25cts be asked.
The motion was lost

This image is from the 1906–1911 record book of the museum (then called simply the Pacific Grove Museum) association advisory committee. The note details their meeting on December 26, the day after Christmas, in 1906. The meeting was held at the museum. Museum curator Mary Norton acted as secretary pro tem for the meeting. (Courtesy Pacific Grove Museum of Natural History.)

Pictured here are the original bylaws for the Pacific Grove Museum Association, as detailed in this handwritten note from 1900. Section 2 of the bylaws reads, "The object of the Association shall be to maintain (1) a Museum established by the Chautauqua Assembly, and to develop a garden of native flora, which together with the Museum shall represent and preserve the life of the region (2) an exchange for providing specimens of the life of the distinct to scientific and other entities." (Courtesy Pacific Grove Museum of Natural History.)

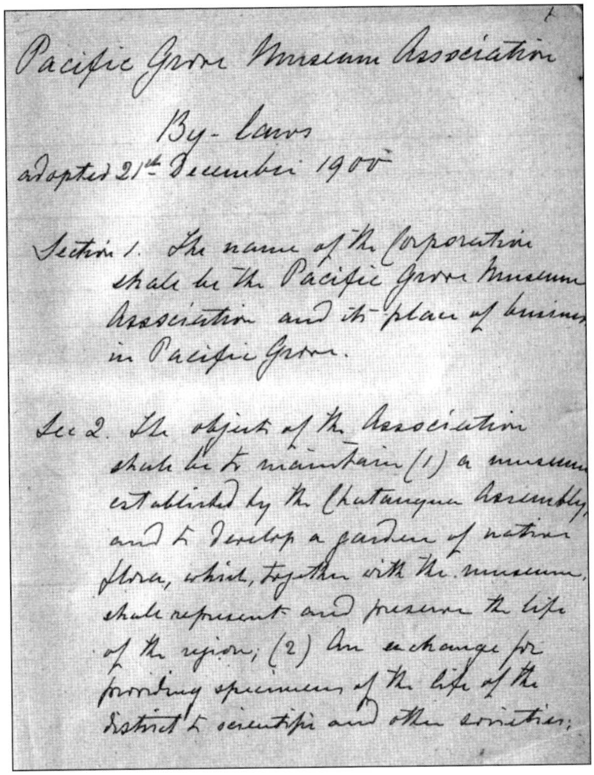

Pacific Grove Museum Association

By-laws

adopted 21st December 1900

Section 1. The name of the Corporation shall be the Pacific Grove Museum Association and its place of business in Pacific Grove.

Sec 2. The object of the Association shall be to maintain (1) a museum established by the Chatauqua Assembly, and to develop a garden of native flora, which, together with the museum, shall represent and preserve the life of the region; (2) An exchange for providing specimens of the life of the District to scientific and other societies;

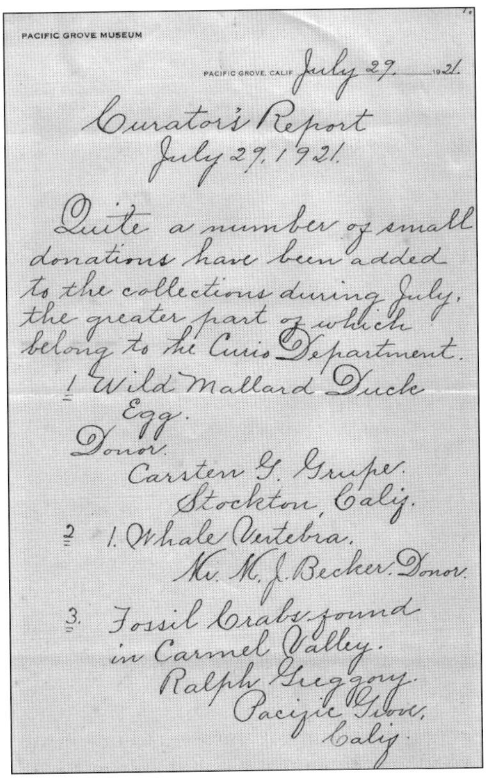

PACIFIC GROVE MUSEUM

PACIFIC GROVE, CALIF. July 29. 1921.

Curator's Report
July 29, 1921.

Quite a number of small donations have been added to the collections during July, the greater part of which belong to the Curio Department.

1 Wild Mallard Duck Egg.
Donor.
Carsten G. Grupe.
Stockton, Calif.

2 1. Whale Vertebra.
Mr. M. J. Becker. Donor.

3. Fossil Crabs found in Carmel Valley.
Ralph Gregory.
Pacific Grove,
Calif.

This is a curators' report from July 29, 1921, that details items donated to the museum collection that month. The report indicates many of the donated items that month were for the museum's curio department. These items included a mallard duck egg, a whale vertebra, and a fossil collected from Carmel Valley. (Courtesy Pacific Grove Museum of Natural History.)

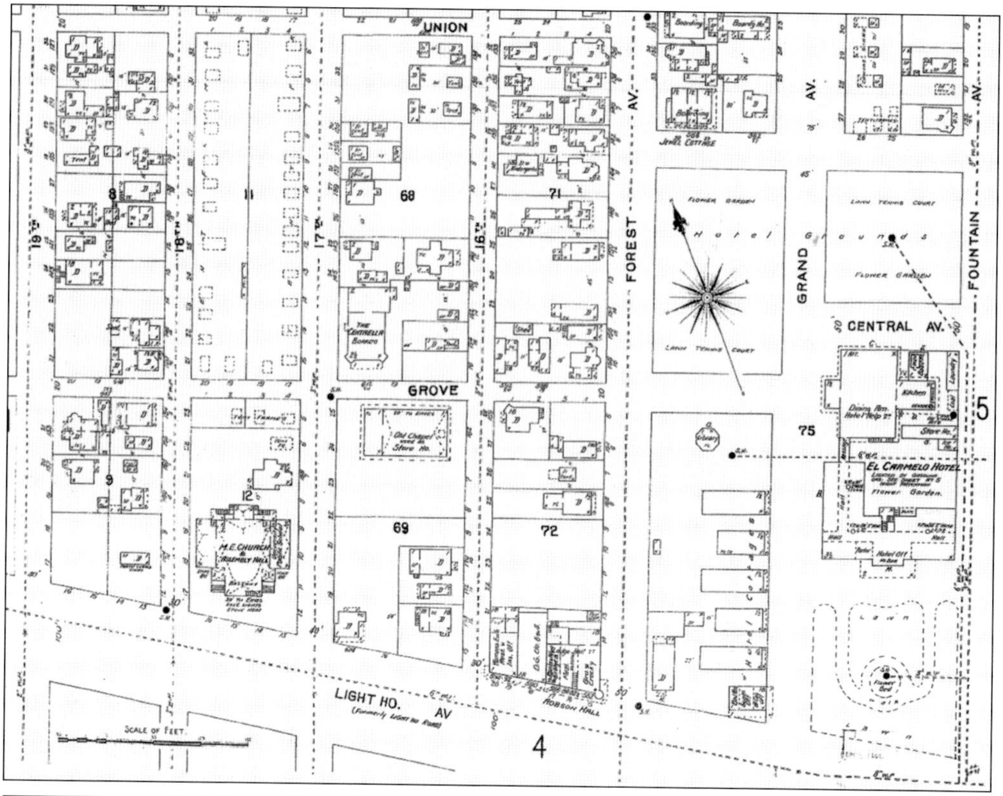

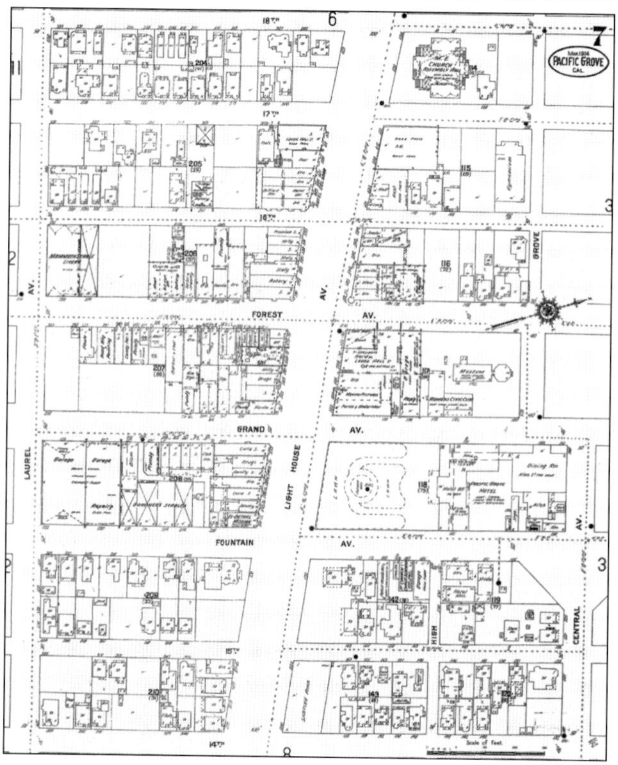

This Sanborn map from 1897 shows the way the museum appeared at that time. It shows the museum located between Forest and Grand Avenues across the street from the El Carmelo Hotel. A number of these Sanborn maps would be made over the years. Each would show different iterations of the streets, and the museum location, which changed slightly over the years. (Courtesy Pacific Grove Museum of Natural History.)

This 1914 Sanborn map shows the museum as it appeared at the time. The museum retains the original structure but has another addition behind it, and the words, "Heat, Stoves, Lights, and Elect," are printed on it. In 1914, the museum was located near the Women's Civics Club, which faced Grand Avenue. (Courtesy Pacific Grove Museum of Natural History.)

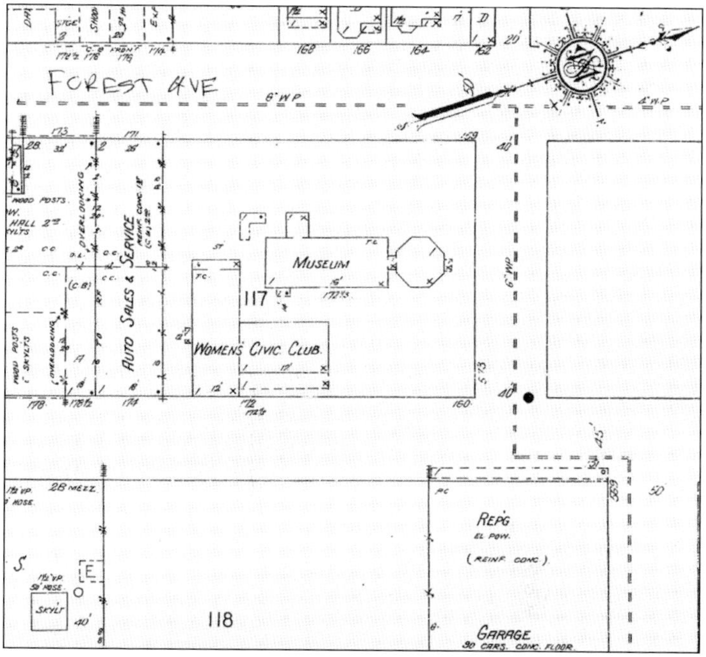

This 1926 Sanborn map shows the configuration of the museum and the property surrounding it as they appeared at the time. The museum would remain in that location, though the building would grow in 1932 and again in 1983 to encompass most of the property, save for a southern area that is home to the Native Plant Garden. (Courtesy Pacific Grove Museum of Natural History.)

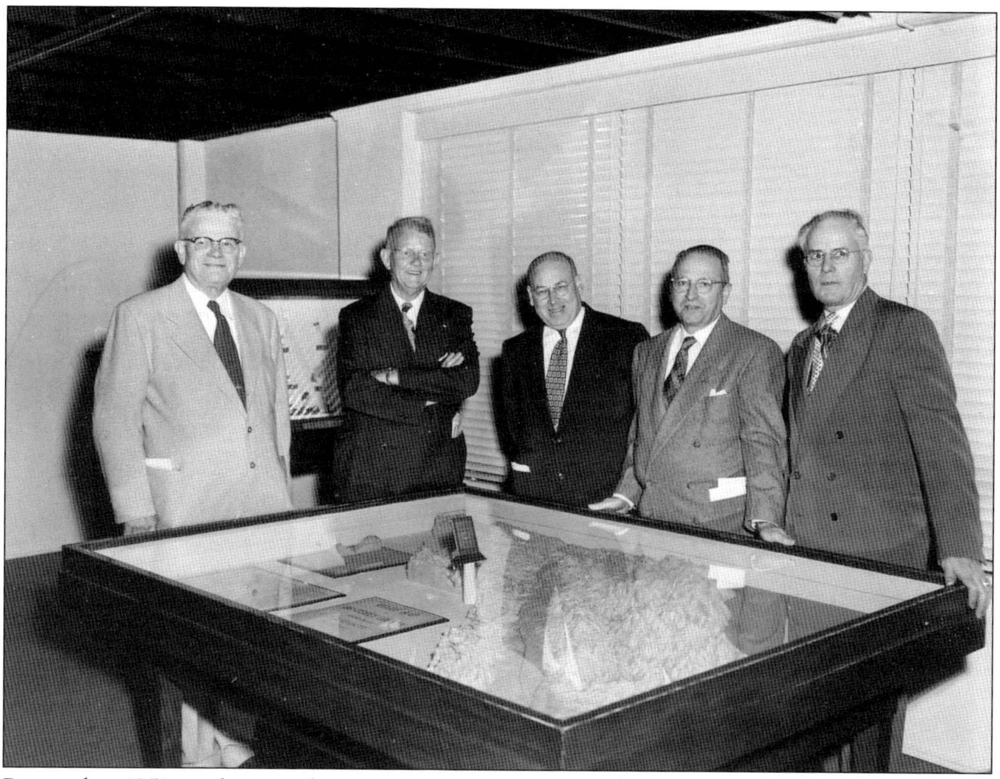

Pictured in 1953 are five members of the museum board as they stand beside a large relief map of the Pacific Ocean shoreline around Monterey County. From left to right are R. Maxwell, ? Rye, ? Lee, ? Hoffman, and ? Dyke, according to museum records. The group is gathered for "Museum Tea." (Courtesy Pacific Grove Museum of Natural History.)

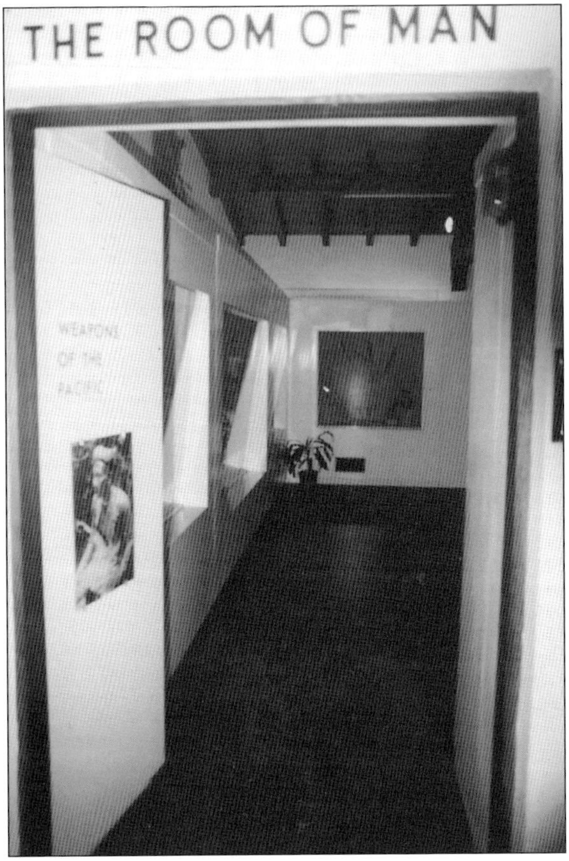

Pictured in 1981 is the entry to the Room of Man, which, at that time, was hosting an exhibit on the weapons of the Pacific. The raised ceiling of the Heritage Gallery can be seen in the photograph, as can a potted plant at the end of the hallway. Three walled, glass exhibit locations are also visible in the image. (Courtesy Pacific Grove Museum of Natural History.)

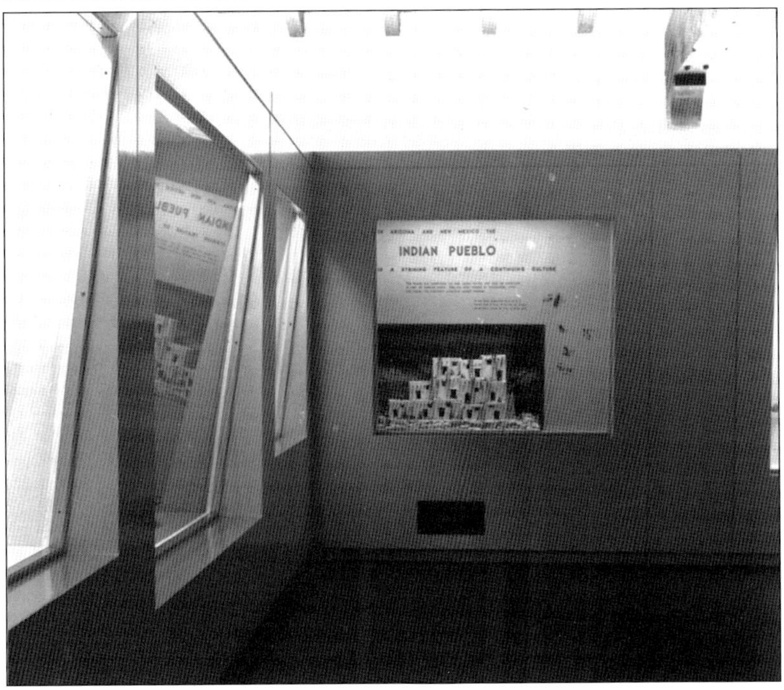

Pictured is the pueblo model as it appeared in 1979 while on public view. The display offered a peek at cultures from both Arizona and New Mexico as part of the exhibit in the Heritage Gallery. The original building beams in the gallery can be seen above the exhibit. (Courtesy Pacific Grove Museum of Natural History.)

A coyote can be seen on the mezzanine level of the museum's Heritage Gallery in this photograph from the early 1980s. The space above the mezzanine is now empty to offer a view of the floor below. The coyote, meanwhile, remains on display in the Heritage Gallery on the first floor. (Courtesy Pacific Grove Museum of Natural History.)

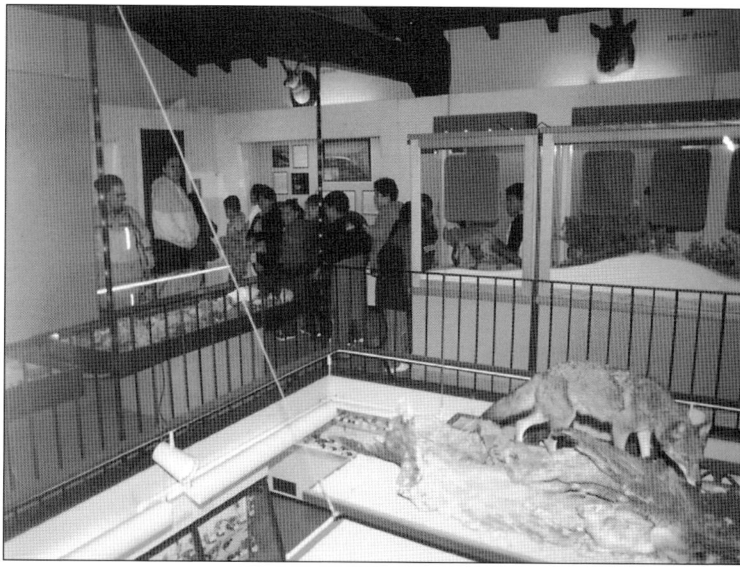

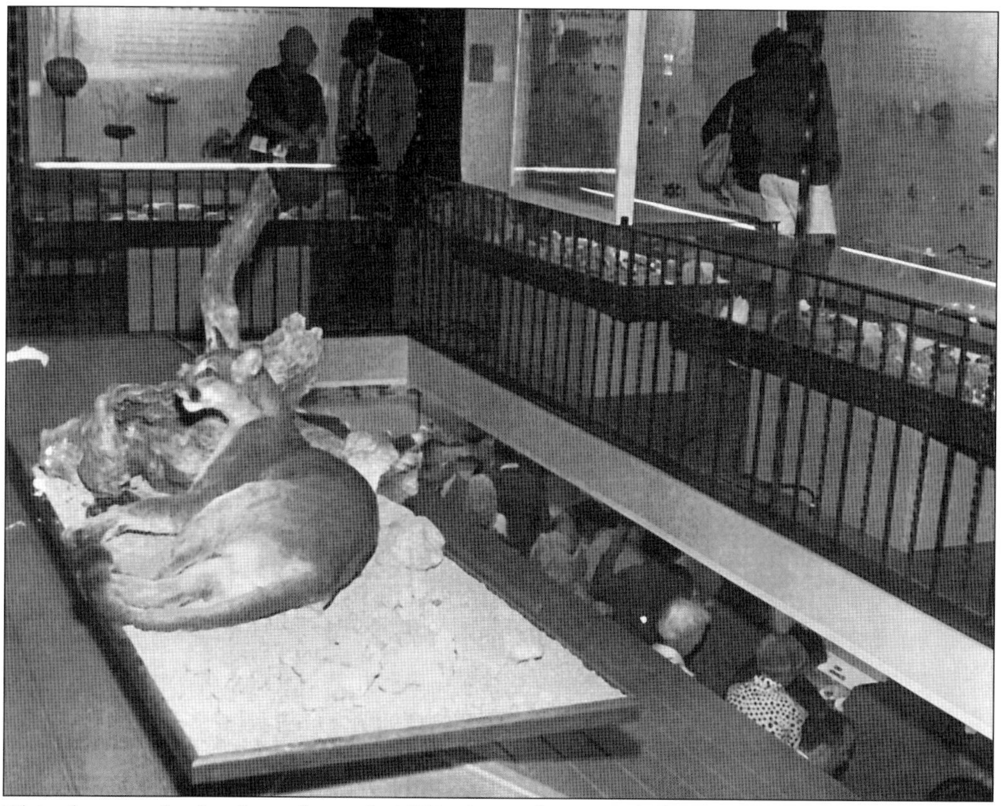

This photograph also from the early 1980s shows the second-floor mezzanine in the museum's Heritage Gallery. Like the coyote, the mountain lion has been moved behind glass on the first floor since this photograph was taken. A large crowd can be seen below the lion as they look at exhibits. (Courtesy Pacific Grove Museum of Natural History.)

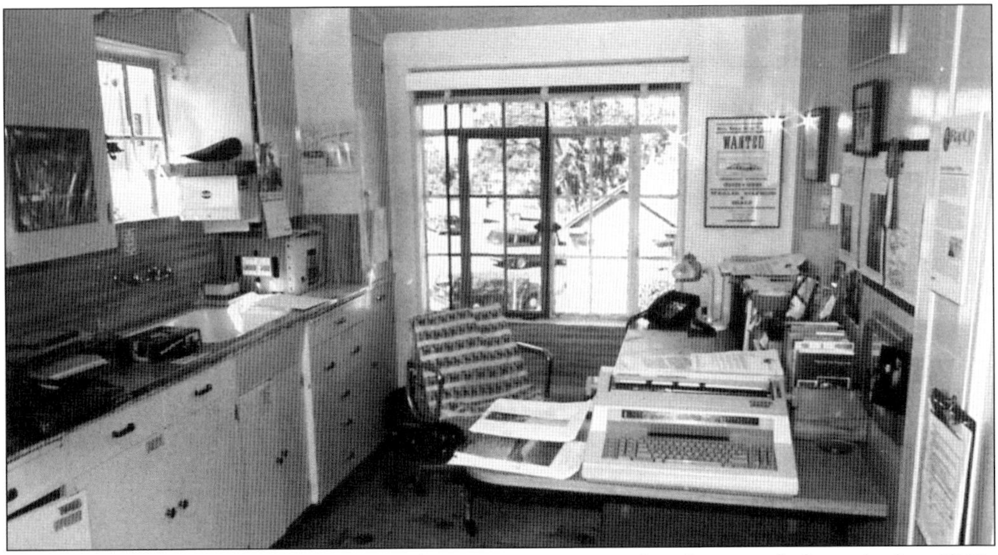

Formerly used as the upstairs apartment for museum curators, this photograph from the late 1980s shows the kitchen area of the apartment. The spaces, built in 1938, feature a bathroom with a shower to this day. The apartment today is used for office space by museum staff and as the main office for the museum's executive director. (Courtesy Pacific Grove Museum of Natural History.)

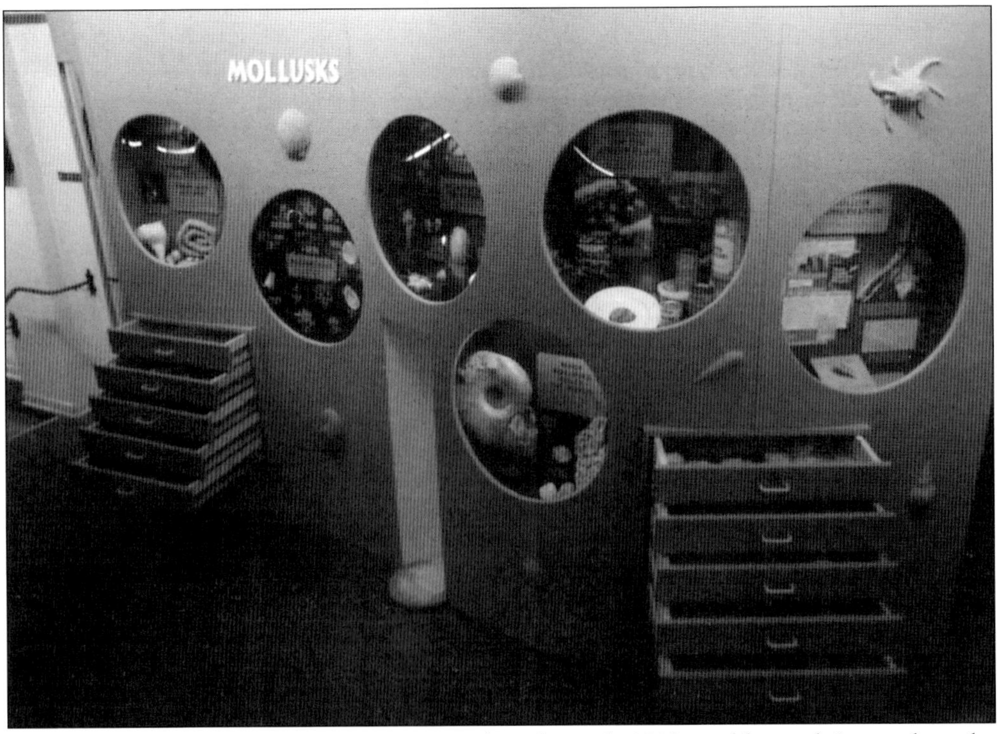

Pictured is the mollusks exhibit as it appeared in the early 1990s and beyond. Located on the second floor of the Heritage Gallery, the exhibit offered information and specimens of shells, squid, and much more. Drawers in the exhibit opened to reveal more information and specimens, including smaller shells. The museum has cabinets full of shells, collected all over the world, in its basement collection. (Courtesy Pacific Grove Museum of Natural History.)

The museum's gift store, pictured here, opened to the public in 1987 by the museum association. It was originally operated by Dorothy Capen and Gertrude Chappell. More than 43 volunteers kept the store operating after its first year. Today, the museum store remains a part of the museum. (Courtesy Pacific Grove Museum of Natural History.)

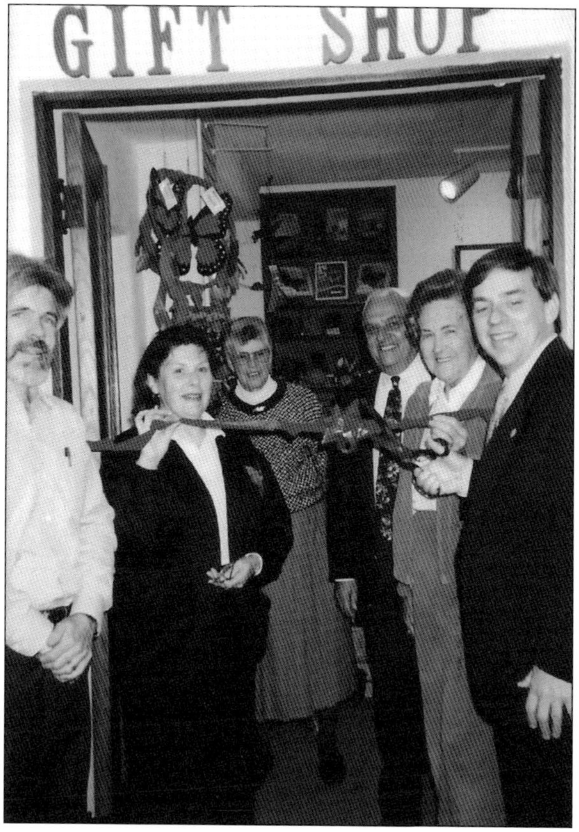

The gift store, pictured here in the late 1980s, offered many books for sale, with many on the native plants of Monterey County. Shells, stuffed animals, rocks, shirts, and much more. According to longtime volunteer Mary Dainton, cashiers in the museum store originally did all the sales by hand. These days, sales are conducted via iPads. (Courtesy Pacific Grove Museum of Natural History.)

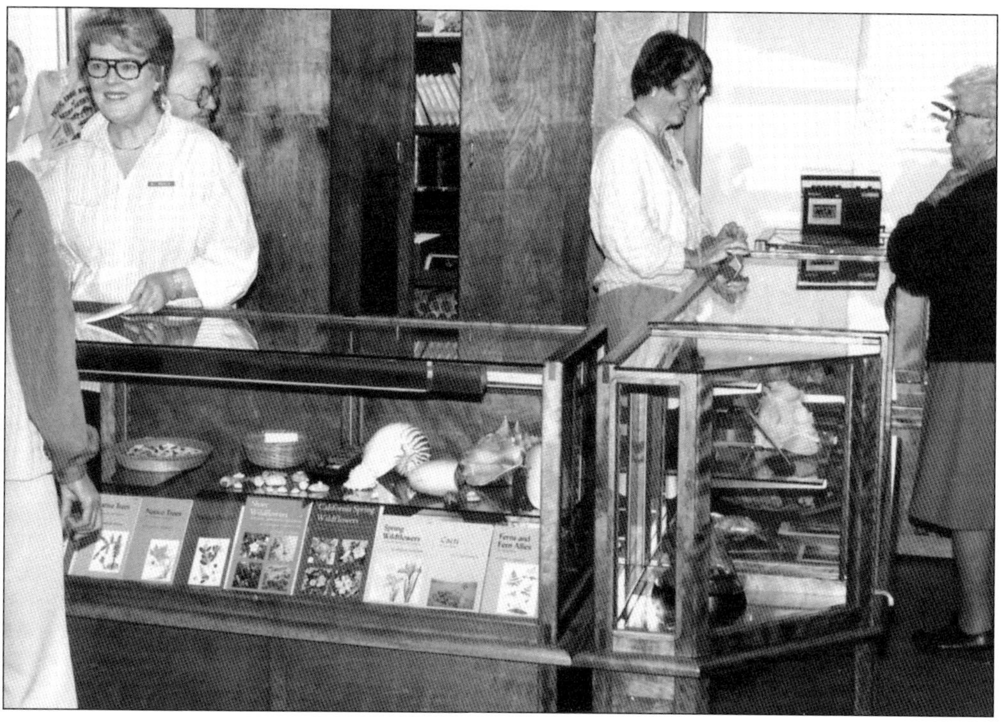

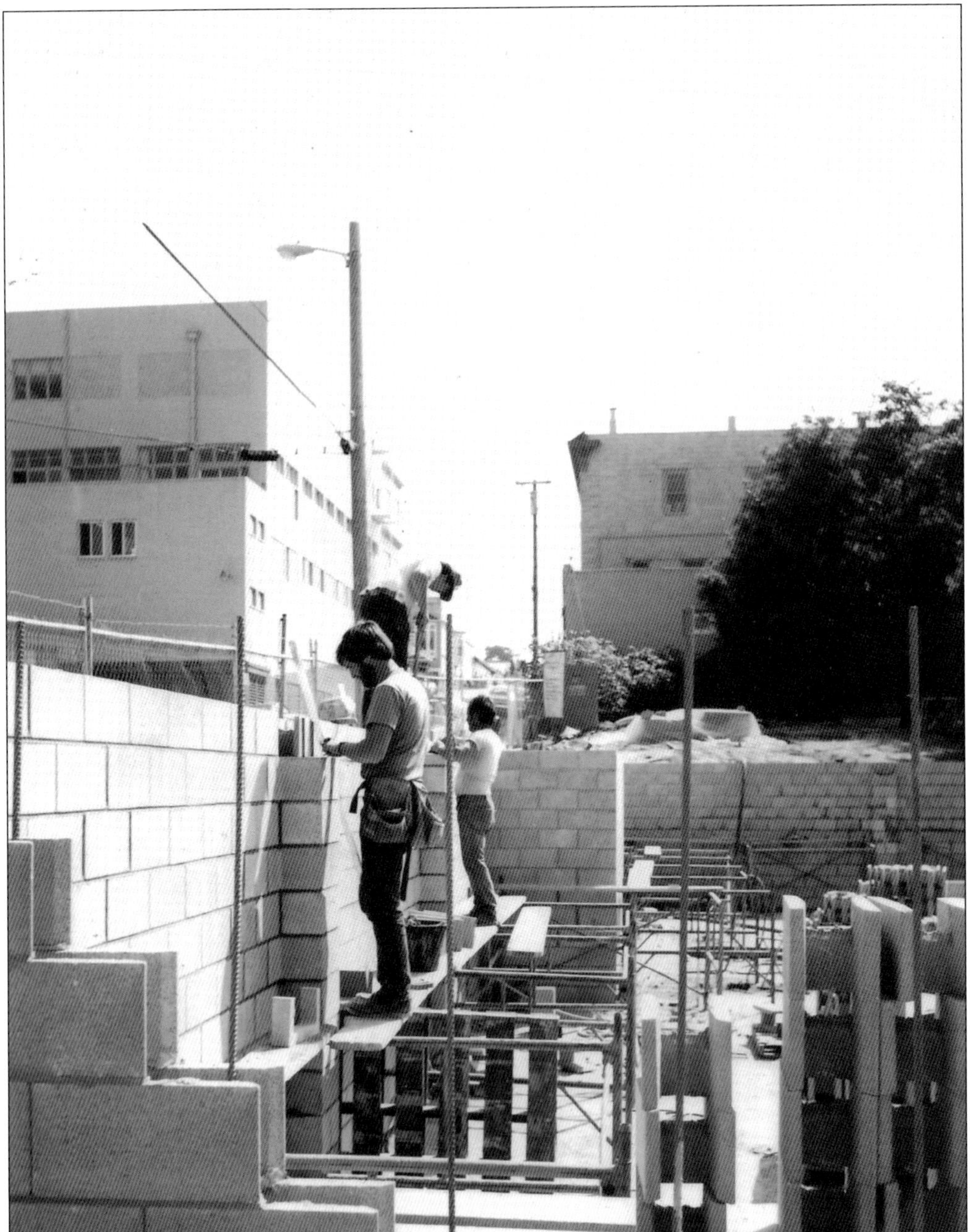

Work on the new museum wing begins. This photograph from 1983 shows crews building the new addition to the museum. The addition more than doubled the size of the old museum building constructed in the 1930s. The Holman building is visible up Grand Avenue in the distance. Ground breaking on the new addition was held in January 1983. Construction cost more than $600,000 and included restrooms, handicap access, additional office space, education offices, a new gallery, and more. The expansion also included the basement, where much of the museum's extensive collection can be found today. (Courtesy Pacific Grove Museum of Natural History.)

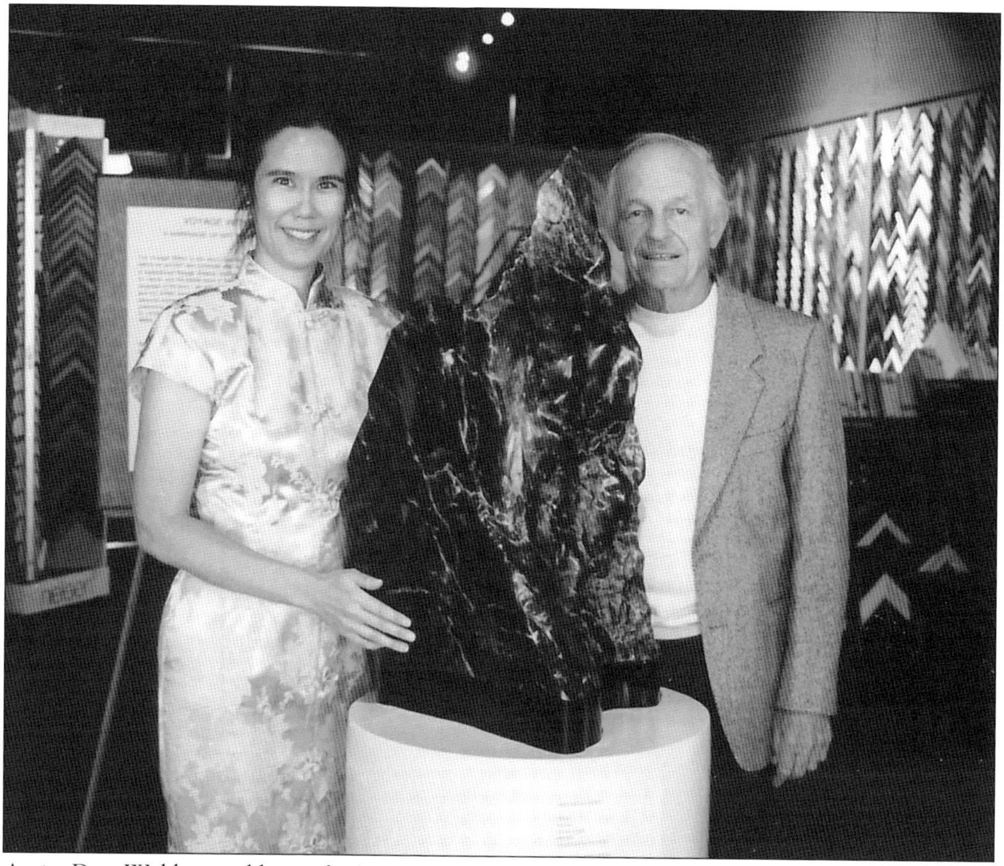

Artist Don Wobber and his wife, Donna, stand by a piece of jade in this 1997 season's greetings photograph. Wobber regularly collected jade, including a sizable piece now found in the museum's Native Plant Garden. The jade boulder, Leucothea, is one of many pieces collected by Don. (Courtesy Pacific Grove Museum of Natural History.)

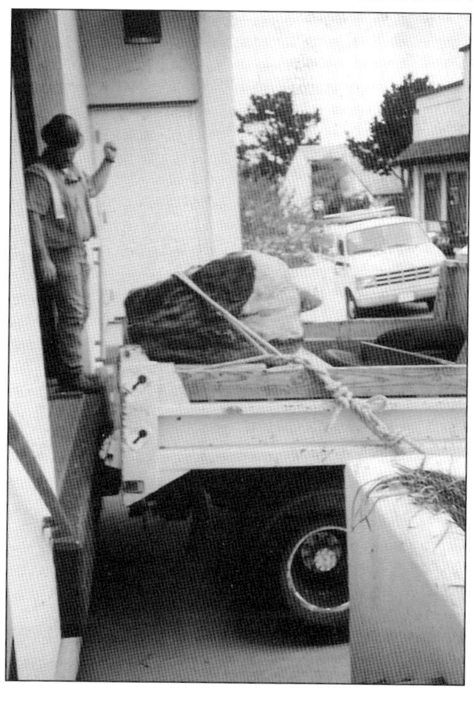

Wobber spent his life collecting nephrite jade from the waters off the Big Sur coastline. The museum's largest piece, Leucothea, is visible in this image from the late 1980s as it was brought by flatbed truck to the museum's Grand Avenue entrance. This rolling door is used for the loading and unloading of large exhibit items. (Courtesy Pacific Grove Museum of Natural History.)

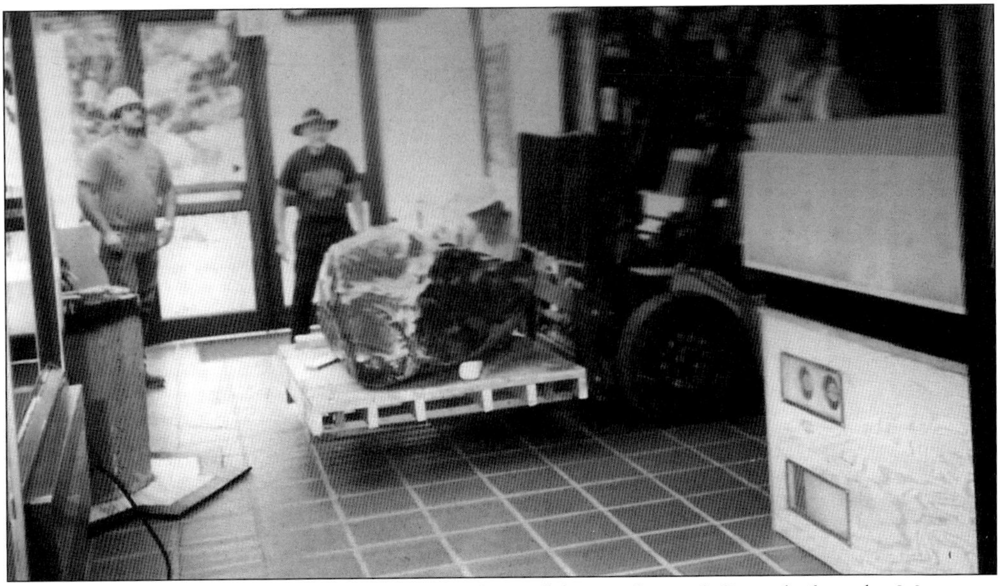

Leucothea was raised from Jade Cove along the California Central Coast before the Monterey area fell under the protection of the Monterey Bay Marine Sanctuary. In this photograph from the late 1980s, a forklift can be seen moving the enormous boulder, sculpted by Wobber, into the museum's new lobby, part of the recent museum expansion. (Courtesy Pacific Grove Museum of Natural History.)

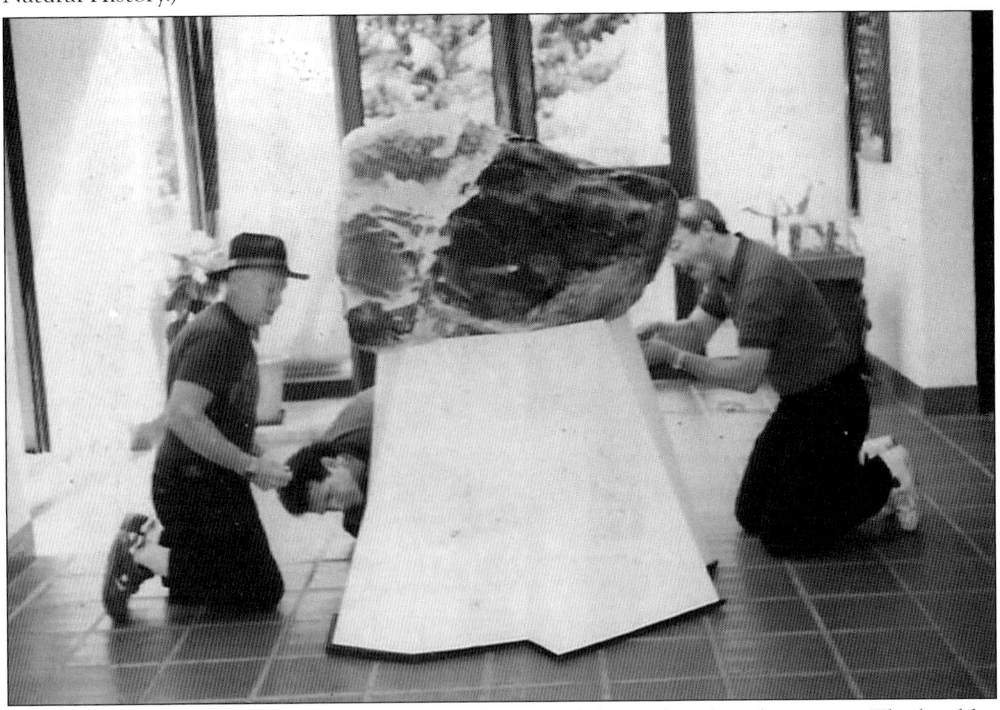

The giant jade boulder, Leucothea, is mounted to an exhibit pedestal in this image. The boulder remained in the lobby for a handful of years before being moved into the Native Plant Garden, where it remains on display to this day. Visitors continue to ask where they can find the large specimen on a weekly basis. (Courtesy Pacific Grove Museum of Natural History.)

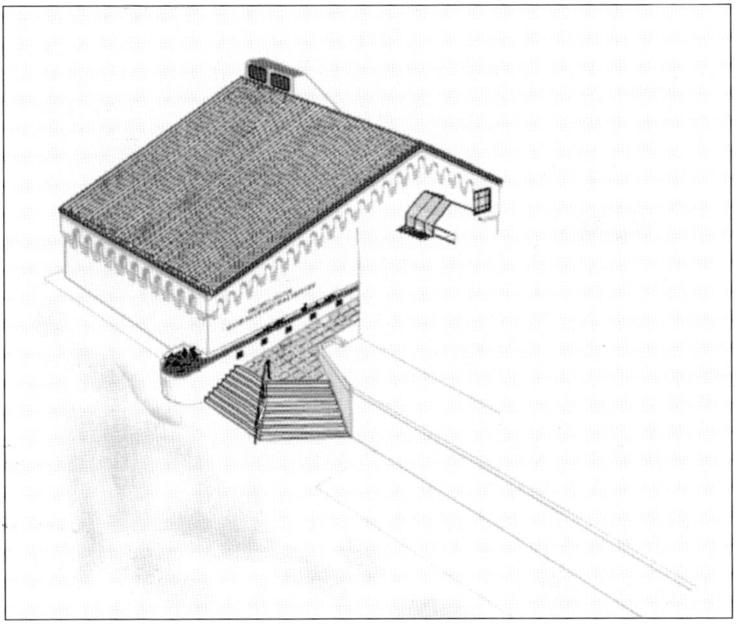

This drawing from Unity Media shows how the expansion to the museum would look when completed. Visible in the image are stairs and handrails leading to a walkway, new signage for the museum that would be visibly from Forest Avenue, the glass skywalk that connects the new addition to the Heritage Gallery on the second floor, and more. (Courtesy Pacific Grove Museum of Natural History.)

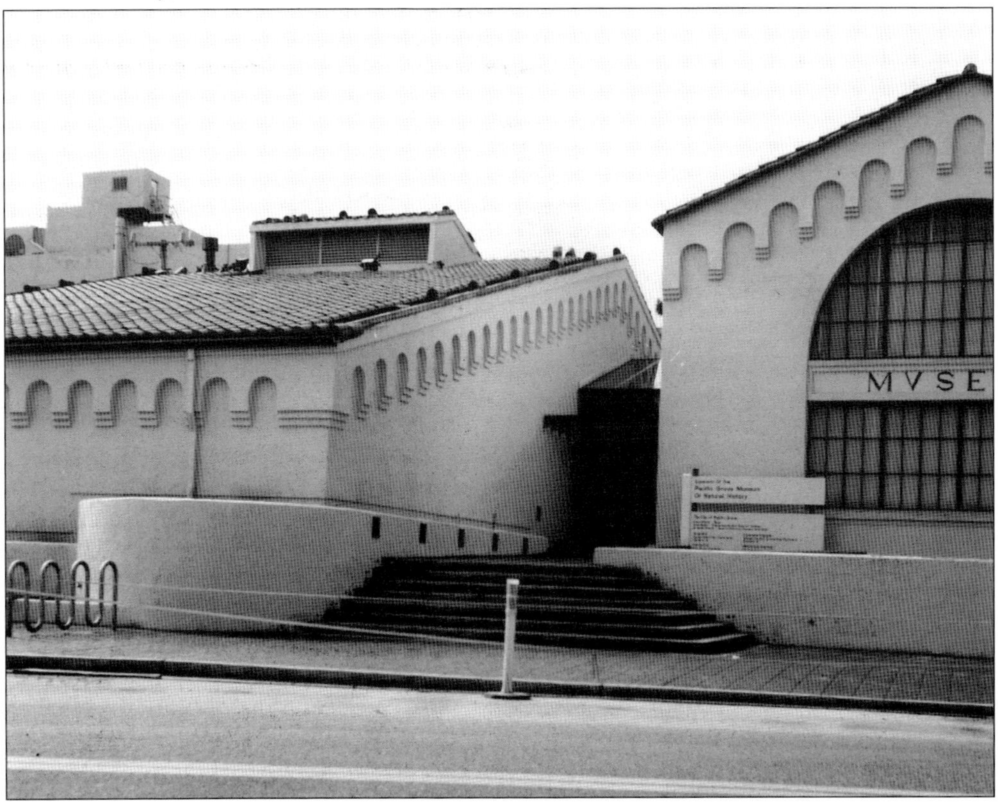

Pictured is how the museum appeared following construction of the new addition in the mid-1980s. The stairs, bicycle rack, and sidewalk in front of the museum are cordoned off. Visible above the roofline of the museum is the Holman building, with a former restaurant space visible at the very top. (Courtesy Pacific Grove Museum of Natural History.)

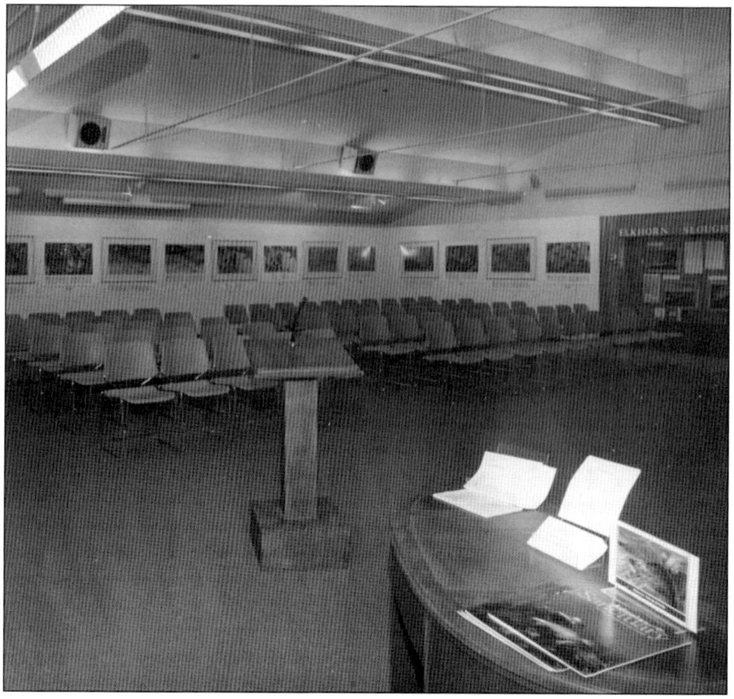

Originally designed as a sort of conference area, the new gallery space created by the 1983 expansion, pictured, made it possible for the museum to hold large meetings on site. The area is now home to the Bird Gallery, though space remains for regular meetings of the Audubon Society, the California Native Plant Society, Sustainable Pacific Grove, the Carmel Valley Gem and Mineral Society, and others. (Courtesy Pacific Grove Museum of Natural History.)

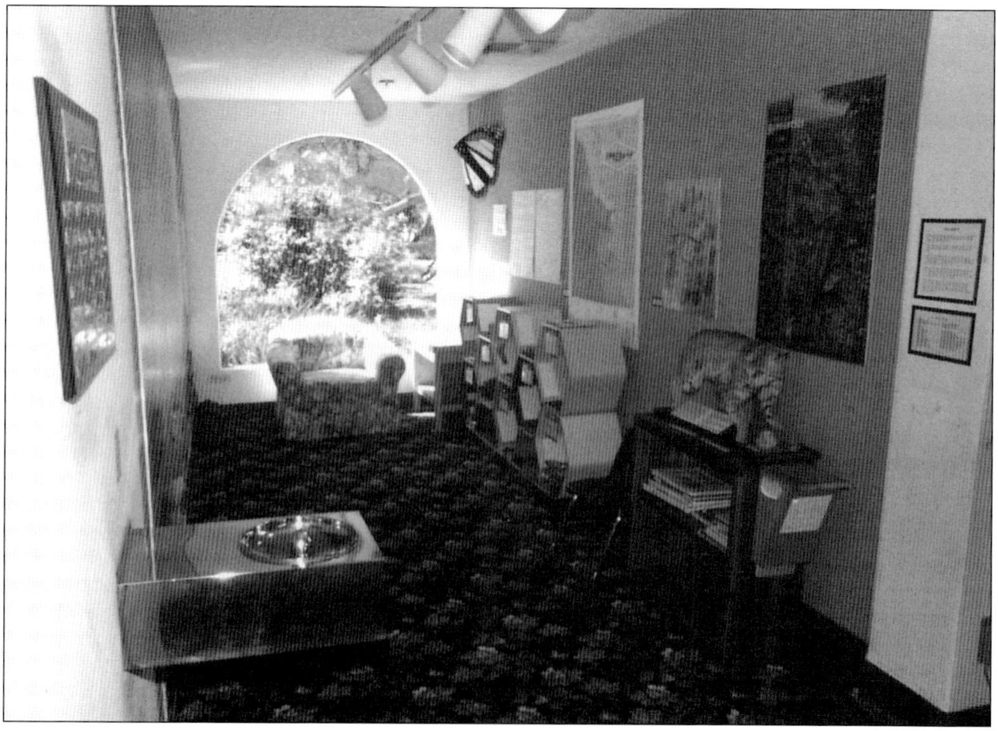

The 1983 expansion also created a space for children to get more hands-on with natural history, as pictured here. A water fountain, cubbies, and a taxidermy cat, as well as a chair and large monarch butterfly, can be seen in this image. A large window at the rear of the space offers a view into the Native Plant Garden. (Courtesy Pacific Grove Museum of Natural History.)

Six

Important Roles

The role of curator is an important one in any museum. From the Latin word *curare*, which means to "take care of," a curator's job has always been to oversee and serve as custodian of a museum collection. More recently, the role of curator has taken on a second role, that of interpretation. Curators often find the best way to explain an object, and place it in context, in the hopes to fully represent the collection to visitors. Beginning with Mary Norton, and under the name Pacific Grove Museum, other curators from 1883 to 1916 included Rev. Thomas Sinex, Laura Duncan, M.E. Hesser, and F.G. Woodcock. After 1916 when the city took control of the museum, Pacific Grove Museum of Natural History curators included Teresa Lloyd, Merton Hinshaw, and Vern Yadon. Hinshaw became curator in 1948 and is remembered most for his craftsmanship with latex. He created many of the items on display at the museum, including the mushroom specimens and many of the lizards and snakes. His skill was sought after by numerous museums across the nation. One museum in Arizona sent him a rattlesnake they hoped to have him make into a mold. What Hinshaw did not expect, however, was for the snake to be alive when it arrived. Other curators who helped shape the current museum include Paul Finnegan, Stephen F. Bailey, and longtime curator Vernal Yadon. Yadon served as curator from 1957 to 1992 and is the museum's curator emeritus. In 2008, the City of Pacific Grove entered into a partnership with the newly formed nonprofit Museum Foundation of Pacific Grove, under the leadership of Lori Mannel, to manage and operate the natural history museum. The city retains ownership of the museum and the collection, while the foundation manages and operates the facility.

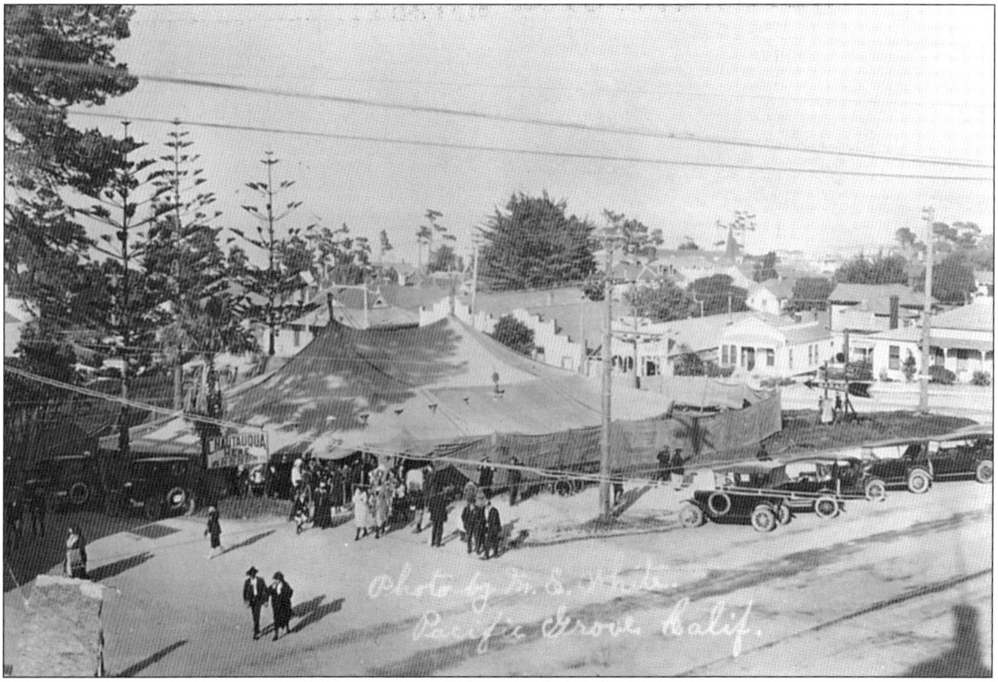

The museum would not be what it is today were it not for the Chautauqua Literary and Scientific Circle, pictured here in 1923 during their last retreat. Retreat staff erected their large tent that year where the Holman building stands today. Construction on the department store started after the tent came down. (Courtesy Layne Collection.)

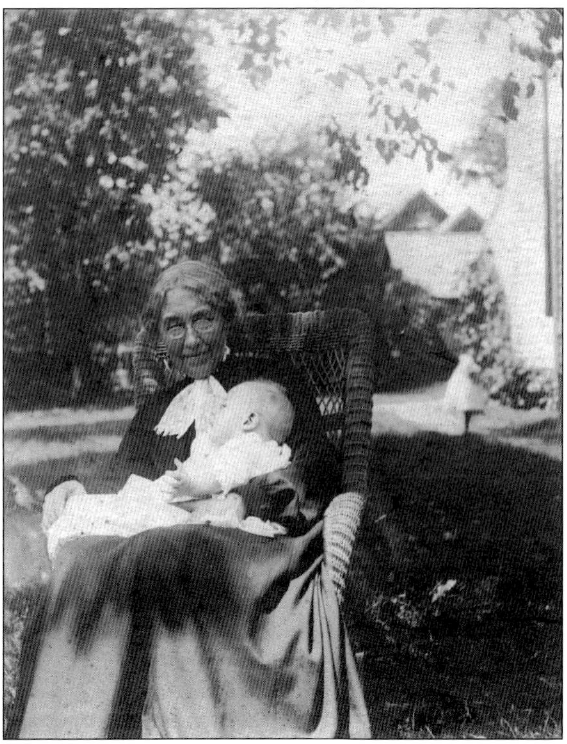

Pictured is Mary E.B. Norton, the botanist founder and first curator of the Pacific Grove Museum of Natural History. Known as "a lady of scholarly attainment and excellent executive ability," Norton taught botany at the Chautauqua Literary and Scientific Circle's Pacific Coast branch. Besides her role as the museum's first curator, this amazing woman also served as curator of the Chautauqua Assembly's scientific collection, housed at a variety of buildings, including Chautauqua Hall, the "old parlor" on Fountain Avenue, and inside the original octagon museum. (Courtesy Pacific Grove Museum of Natural History.)

Museum curator Teresa Lloyd and her daughter Myfanwy smile for the camera in this photograph from the 1930s. The mother and daughter team were the first to live in the rooms built in 1938 above the museum. Myfanwy Lloyd would go on to become assistant curator at the museum. (Courtesy Pacific Grove Museum of Natural History.)

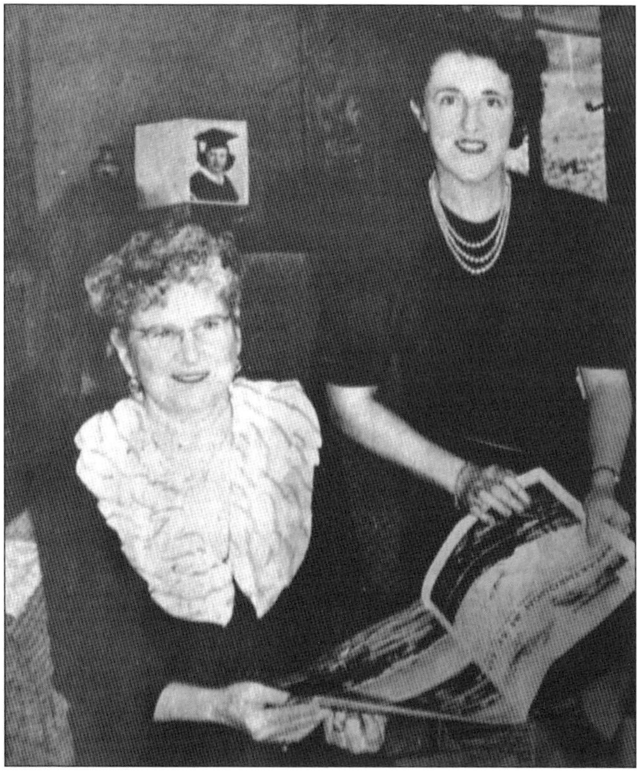

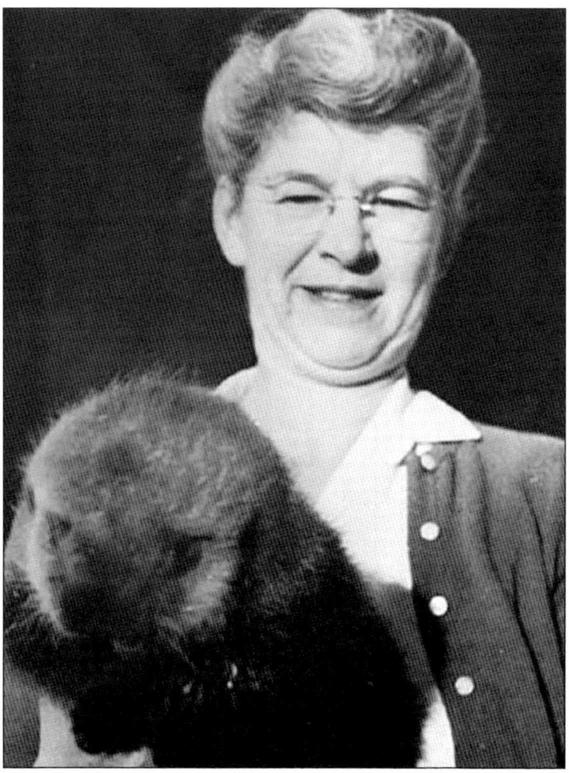

Pictured in the 1930s is assistant museum curator Helen Lind. She is pictured caring for a young sea otter on the museum grounds. The otter was named Bessie Lou during her short stay. She was later taken to the Steinhart aquarium but died 10 days after her arrival there. (Courtesy Pacific Grove Museum of Natural History.)

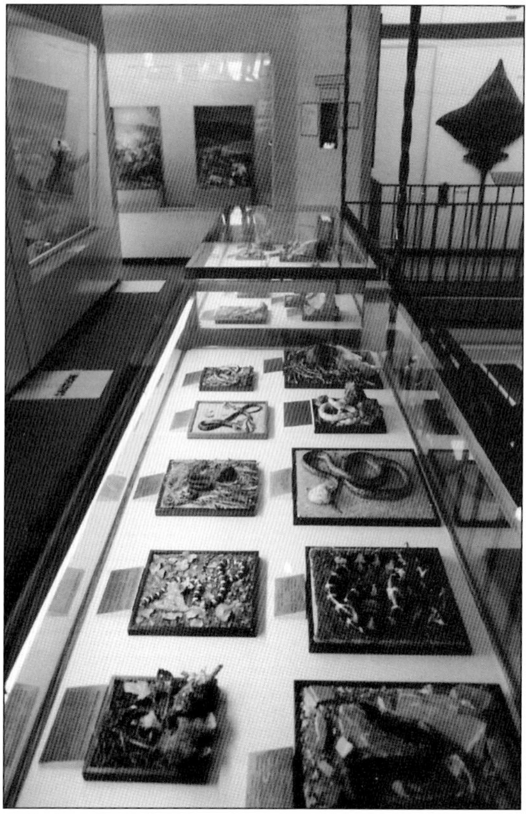

The reptile case is visible in this image from the early 1990s on the mezzanine level of the Heritage Gallery. Many of the reptiles on display there are replicas created by curator Merton Hinshaw. Hinshaw also created molds of mushrooms, many of which are still on display at the museum. (Courtesy Pacific Grove Museum of Natural History.)

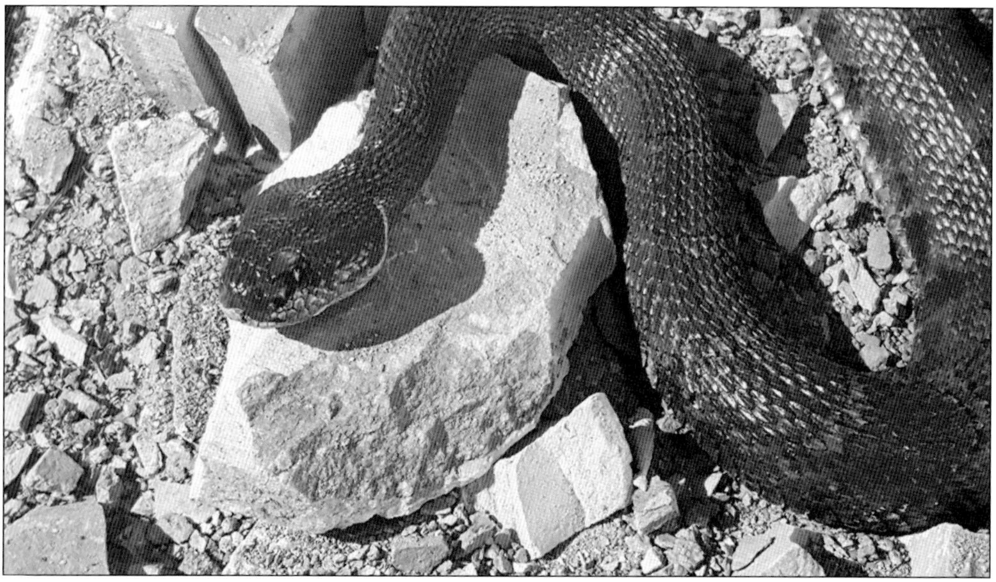

This is an up-close view of a Pacific rattlesnake, the only poisonous snake found in Monterey County. This mold was one of many created by Merton Hinshaw. The reptile case is a favorite among museum visitors. As with most of the items on display in the museum, the reptiles found in the exhibit can all be seen in the wild around Monterey County. (Courtesy Pacific Grove Museum of Natural History.)

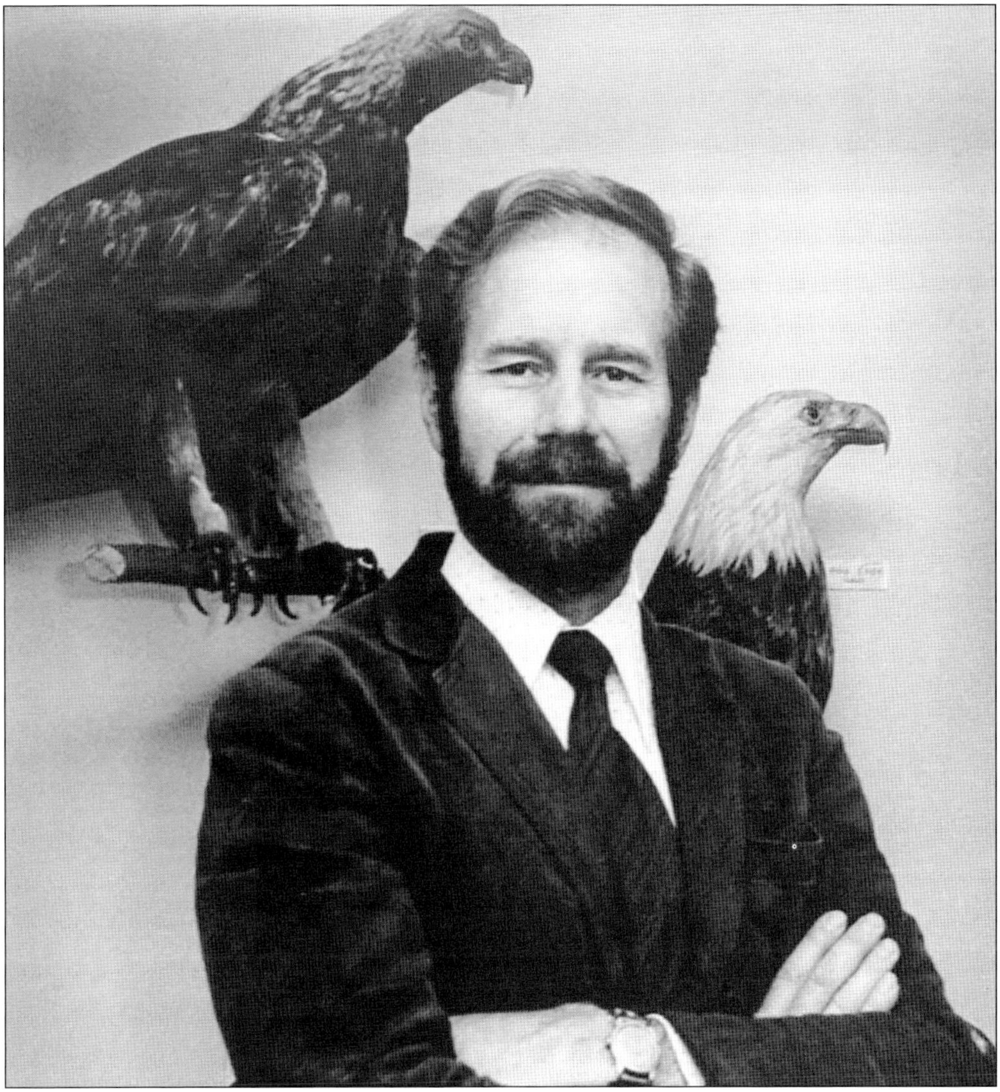

Vern Yadon is pictured in the early 1970s posing with his arms crossed while two birds are visible behind him. Over the years of his tenure as both a city employee and the museum curator, now the museum's director and curator emeritus, Yadon oversaw the museum's growth as well as its accreditation by the American Association in 1972. To this day, Yadon can be found adding plant specimens to the museum's basement collection. His passion for native plants is well known throughout Monterey County, particularly among members of the California Native Plant Society's Monterey Bay chapter. (Courtesy Pacific Grove Museum of Natural History.)

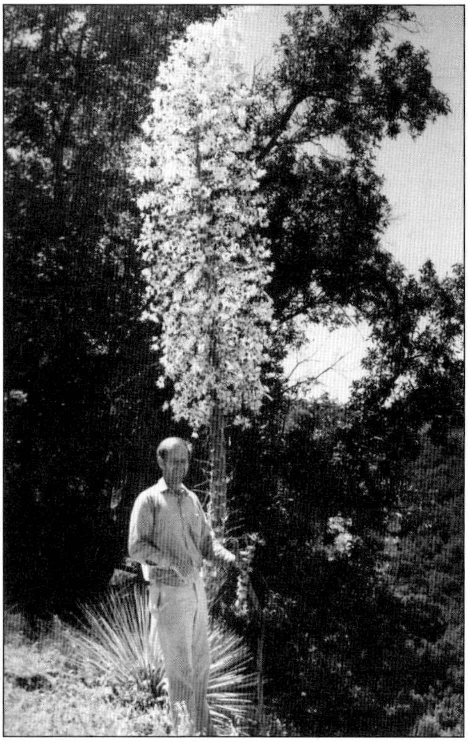

Curator emeritus Vern Yadon is pictured in 1975 standing beside a 16-foot yucca on the Los Padres National Forest. His notes indicate that the yucca had a three-foot diameter. He also mentioned how terrible the deerflies were at the time the photograph was taken. Yadon would serve as curator of the museum for 35 years. (Courtesy Pacific Grove Museum of Natural History.)

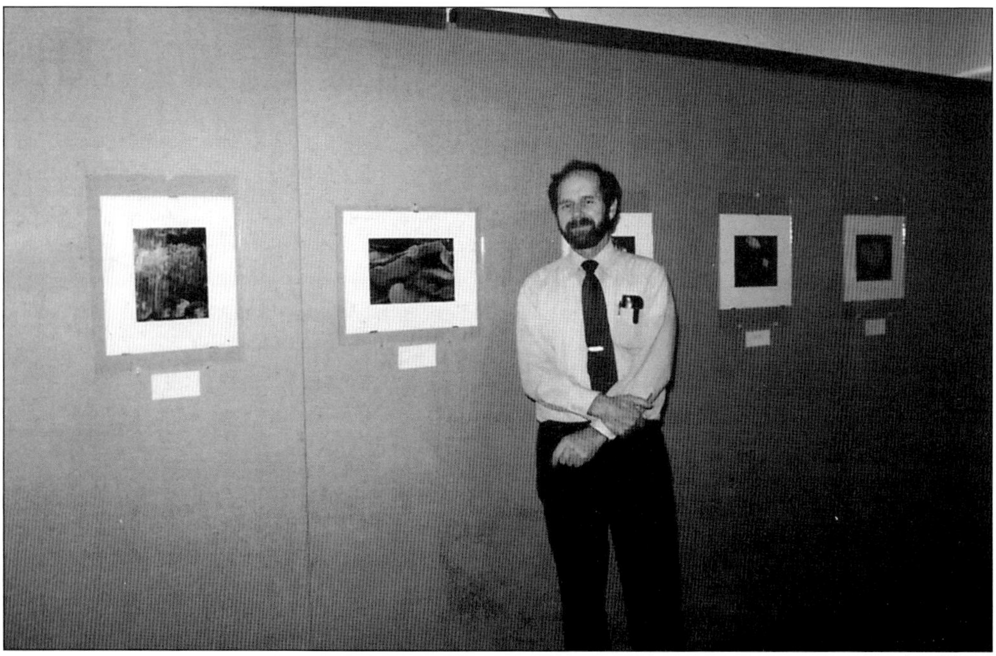

Curator and director Vern Yadon is pictured in 1985 standing before an exhibit of images. Besides setting up exhibits and managing collections, Yadon also worked to acquire new items for the museum, including pottery, many herbarium specimens, and much more. Yadon was also instrumental in developing the Native Plant Garden located at the museum. (Courtesy Pacific Grove Museum of Natural History.)

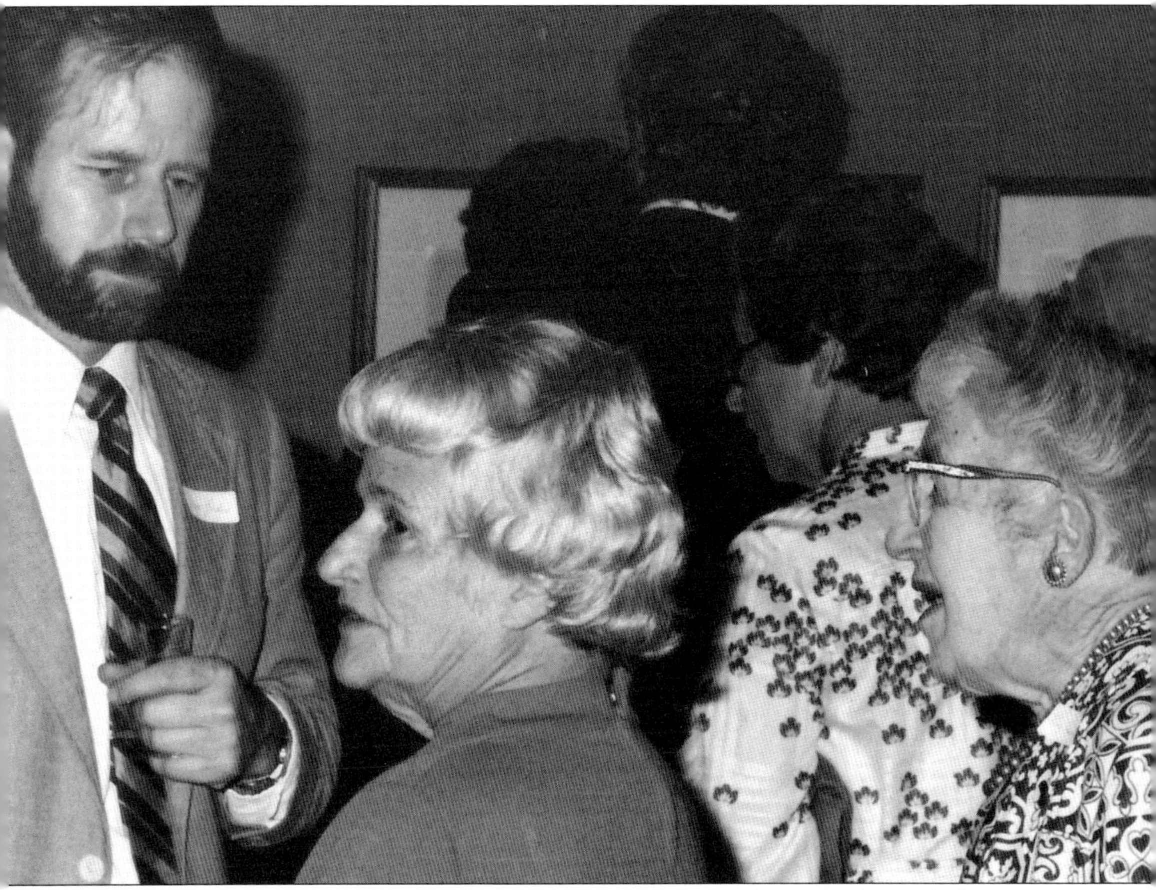

Curator Vern Yadon is pictured here as he talks to guests in September 1978. A crowd of museum visitors is visible behind him and in front of him as they observe a display. The crowds were likely pictured attending an exhibit opening, as evidenced by Yadon's name tag. (Courtesy Pacific Grove Museum of Natural History.)

This is the old museum logo as it appeared in the 1990s. The logo was created in the 1994 to symbolize deep water layers to mountains in its wavy lines. The logo would eventually change to its current image of a monarch butterfly on the letters of the museum's name. Plans began in 2017 for a new logo to reflect the museum's strategic plan. (Courtesy Pacific Grove Museum of Natural History.)

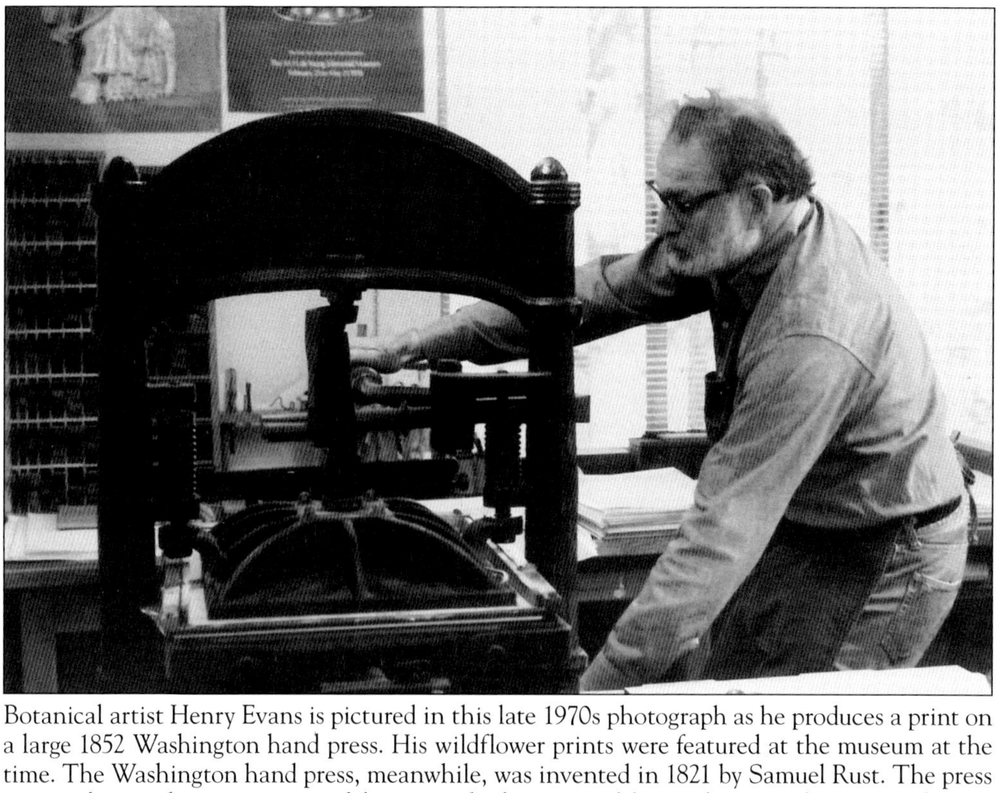

Botanical artist Henry Evans is pictured in this late 1970s photograph as he produces a print on a large 1852 Washington hand press. His wildflower prints were featured at the museum at the time. The Washington hand press, meanwhile, was invented in 1821 by Samuel Rust. The press was similar to other iron presses of the time, which were used for graphic reproductions and more. (Courtesy Pacific Grove Museum of Natural History.)

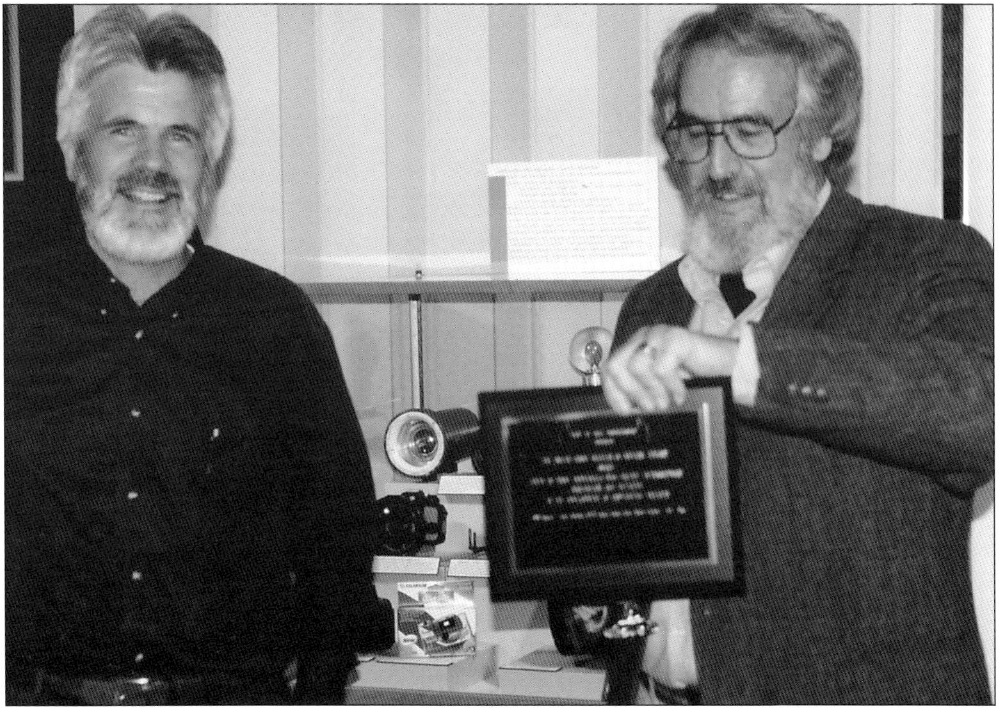

This photograph from 1986 shows Paul Finnegan (left) and Glenn Pollock receiving the "Got It All Together Award." Both appear to be amused by the honor as they stand in the conference room at the museum, an area today known as the Bird Gallery. In the background are accordion screens, behind which the museum stores chairs for meetings and events. (Courtesy Pacific Grove Museum of Natural History.)

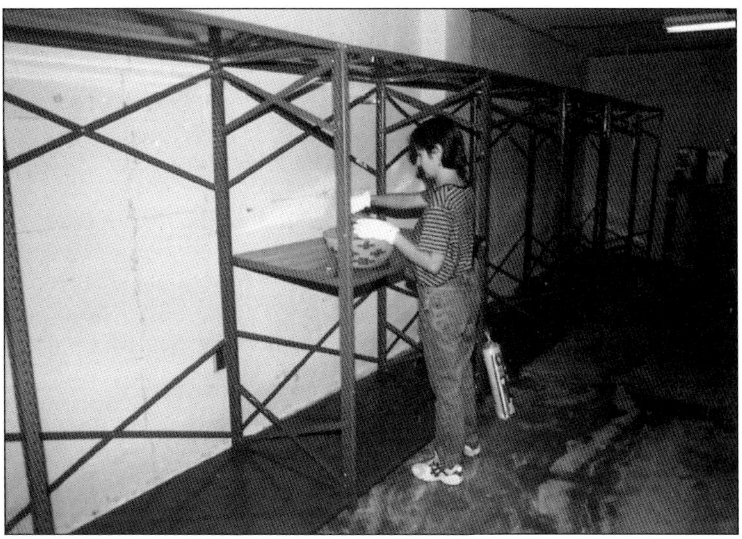

Pacific Grove volunteer Kim Farmer is pictured in the late 1980s as she inventories and rearranges more than 500 baskets in the museum basement. Only a fraction of the museum's collection is on display to visitor. Many more reside beneath their feet in the basement. (Courtesy Pacific Grove Museum of Natural History.)

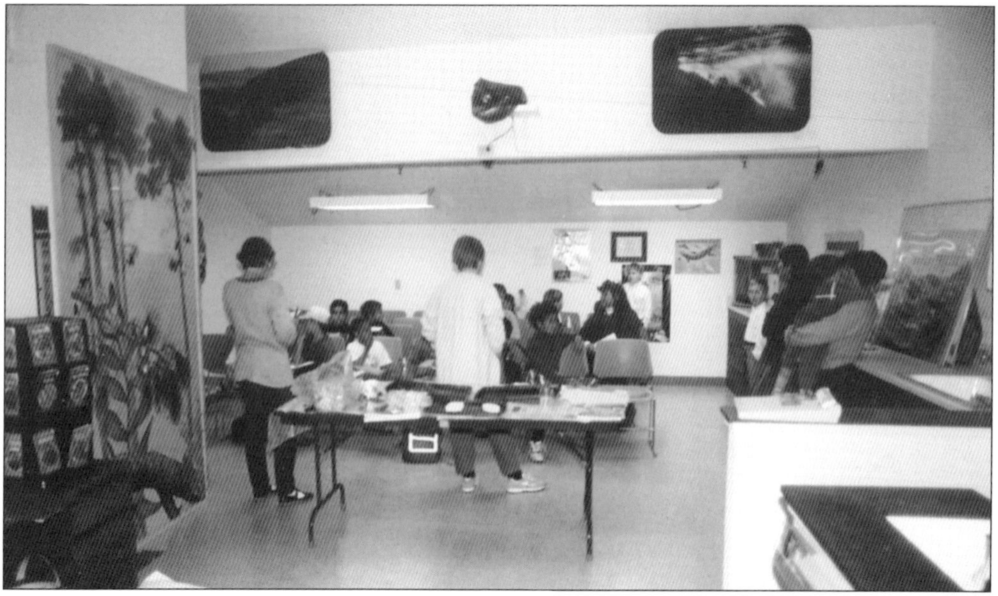

The Education Room is pictured here in the late 1980s, according to museum records. Students can be seen seated at the back of the room, while educators stand beside tables in front of them. Also visible are an assortment of educational materials. The room is now used as offices for the museum's growing education staff. (Courtesy Pacific Grove Museum of Natural History.)

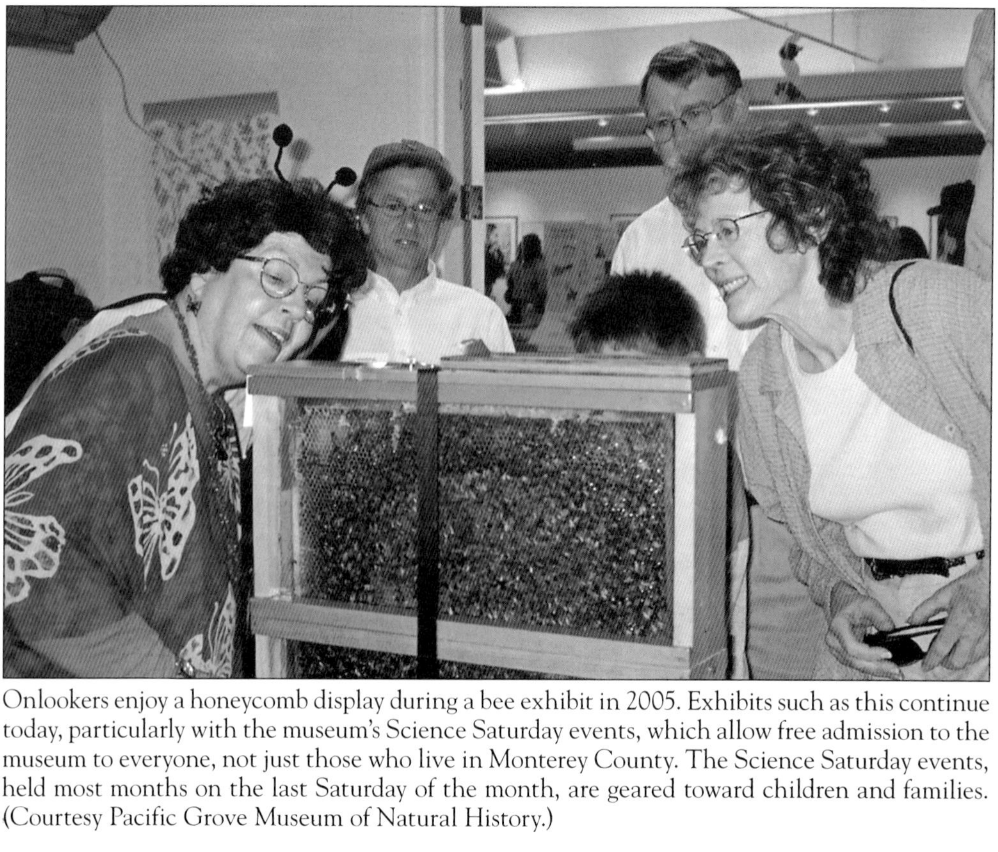

Onlookers enjoy a honeycomb display during a bee exhibit in 2005. Exhibits such as this continue today, particularly with the museum's Science Saturday events, which allow free admission to the museum to everyone, not just those who live in Monterey County. The Science Saturday events, held most months on the last Saturday of the month, are geared toward children and families. (Courtesy Pacific Grove Museum of Natural History.)

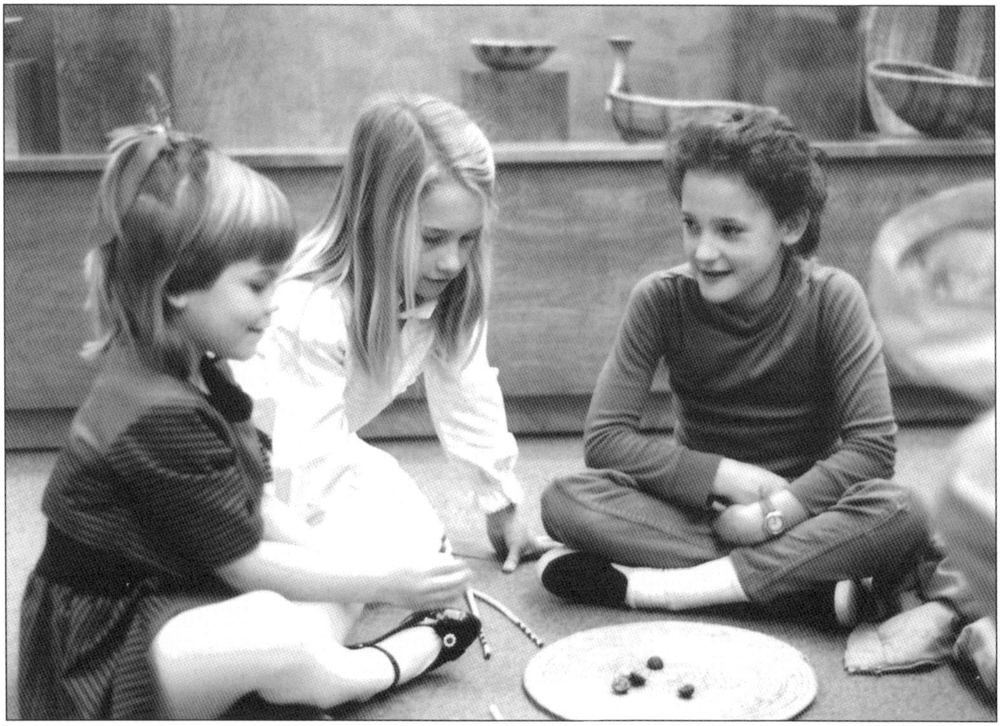

A group of children can be seen participating in group activities at the museum in 1989. Often, the activities for children are designed around current exhibits displayed at the museum. These three children are playing with a circular woven object and assorted pieces while surrounded by Native American baskets. (Courtesy Pacific Grove Museum of Natural History.)

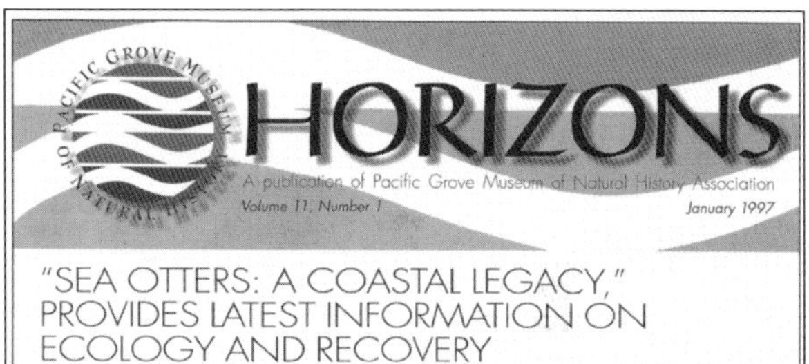

PACIFIC GROVE MUSEUM OF NATURAL HISTORY

HORIZONS

A publication of Pacific Grove Museum of Natural History Association

Volume 11, Number 1

January 1997

"SEA OTTERS: A COASTAL LEGACY," PROVIDES LATEST INFORMATION ON ECOLOGY AND RECOVERY

Friends of the Sea Otter (FSO) and the Pacific Grove Museum of Natural History have joined together to create a long-running exhibition, "Sea Otters: A Coastal Legacy" on display from January 25 through May 11. It will provide visitors with the unique opportunity to see the wealth of information currently available on this popular marine mammal.

Sea Otters were hunted nearly to extinction from 1750 to 1911 for their lush fur, far more dense than any other animal's. In fact, the California Sea Otter was believed extinct until a small band of survivors were "rediscovered" when the Pacific Coast Highway was pushed through the rugged Big Sur area.

"FSO was founded in 1968 in response to the need to protect the small, struggling population of southern (California) Sea Otters," says Susan Brown, Executive Director. "In 1977, we were instrumental in getting this remarkable animal listed as a threatened species under the Endangered Species Act."

Not every person welcomes the return of the Sea Otter. They do consume great quantities of shellfish, some of which are sought by humans as well. Moreover, there is disagreement over whether the current Califor-

(Continued on page 6)

Mother Sea Otter holding her 4-day old pup, a female, in Seattle Aquarium in 1980. The pup has just nursed, has turned about, and now has its head near mother's chin. Victor B. Scheffer photo

PENINSULA INSTITUTIONS HELP CELEBRATE 'WHALEFEST'

The Gray Whales are coming back to California, and the Monterey Peninsula will greet them with "Whalefest," a two month celebration that features whale-watching, a film festival, special exhibits, children's activities, art shows, living history and other events.

"Whalefest" is the first collaborative event by Monterey Association of Cultural Institutions (MACI). Scheduled from January 17 through March 16, it is designed to attract offseason visitors.

Monterey Bay Aquarium will kick off the celebration with a program to provide attendees with whale-related activities. Other institutions including this Museum will schedule educational and entertaining whale displays and exhibits throughout the two-month period.

(Continued on page 7)

One of Lou Silva's murals of whales graces a San Francisco wall. A renowned artist, Silva's drawings and renderings will be featured at the Museum as part of Whalefest.

Pictured is the *Horizons* museum newsletter from January 1997. The newsletter was produced by members of the museum association every January, April, July, and October. The quarterly publication ended with a special commemorative issue in 2010. Editions of the newsletter were sent to members by mail upon publication. *Horizons* coeditors included Sharon Blaziek and Nancy York. (Courtesy Pacific Grove Museum of Natural History.)

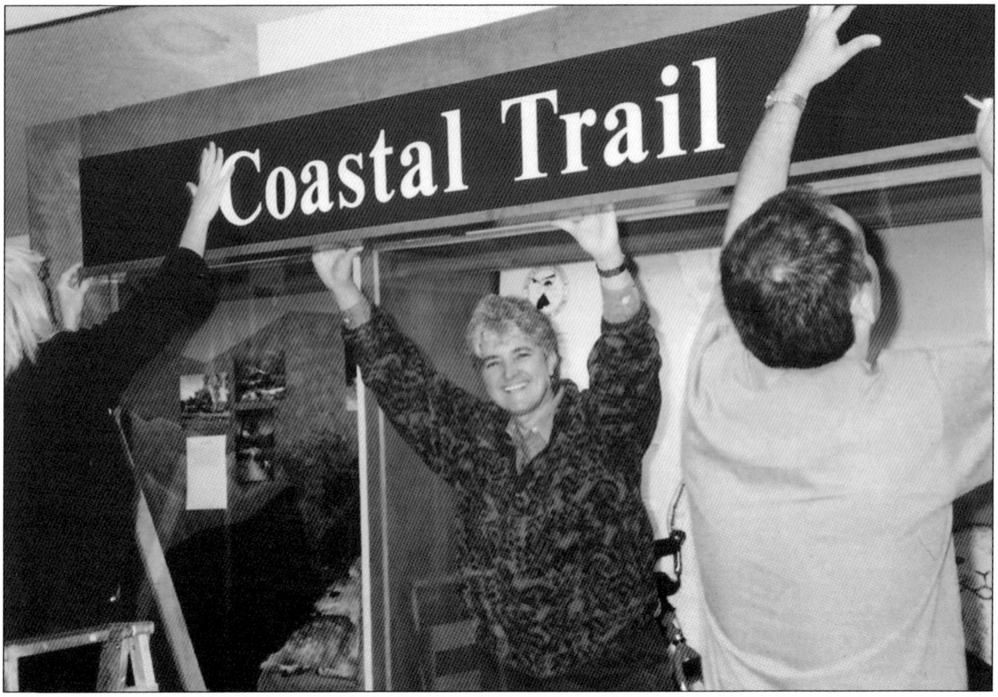

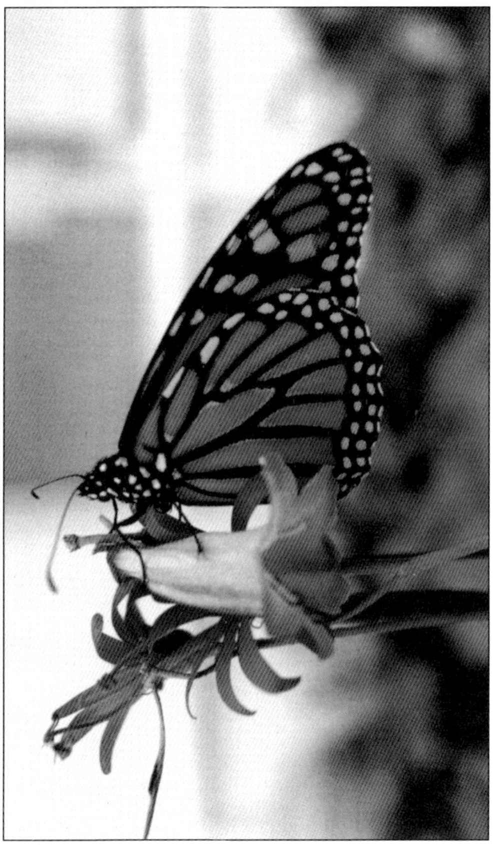

Longtime volunteer Mary Dainton is pictured in 1998 helping set up an exhibit display. Dainton has volunteered at the museum in many capacities, including as a docent at the Monarch Grove Sanctuary, where the monarch butterflies overwinter each year, and in the museum store, as well as helping host special events. (Courtesy Pacific Grove Museum of Natural History.)

This 1993 postcard celebrates the gala opening of the museum's October 9 exhibit opening, A Multiplicity of Monarchs. Photographer Sally Herrgott provided the magnificent photograph used for the postcard. The exhibit, which featured lectures and activities, ran until December 5, 1993. The gala opening began directly following city's monarch butterfly parade. (Courtesy Pacific Grove Museum of Natural History.)

Seven

Sandy, Dinosaurs, and Beyond

The story of the Pacific Grove Museum would not be complete without mention of one of its most famous features. Sandy the Whale has inhabited the area outside of the museum since the mid-1980s and is one of the city's most iconic features. The 40-foot female gray whale, made of segmented ferro-cement, greets visitors daily. She is part of not just the museum's story, but Pacific Grove's story, and has a place in the family photo albums of many residents and visitors. Sandy's fascinating existence did not start at the museum, however. The brainchild of artist Larry Foster, Sandy got her start in an Oakland art studio in 1971. Foster created eight segmented sections that, when put together, form Sandy's 40-foot, 6,000-pound body. In November 1974, Sandy debuted in Pasadena at the California Institute of Technology's Baxter Gallery as part of their exposition on whales. She would go on to travel across the country as part of Foster's General Whale company, which fit in well during a period marked for its efforts to save the whales. Sandy initially arrived in Pacific Grove for a three month stay in November 1980, where she was welcomed by Pacific Grove mayor Florus Williams and museum curator Vern Yadon. Sandy's stay was extended to August due to her popularity, however, and Foster soon offered to sell her to the museum for $24,000. The community rallied behind the effort and, by late 1982, raised a whopping $31,000 to purchase the five-ton sculpture. The funds also paid to move Sandy from her location on Forest Avenue to her current location, "Sandy Plaza," on Central Avenue. While Sandy was the first large visitor to the museum, she was followed a few years later by seven larger than life guests. In June 1989, the museum played host to the Dinosaurs Alive! exhibit, which featured a triceratops, tyrannosaurus, pteranodon, parasaurolophus, a brontosaurus, and two baby "bronties." The creatures moved and roared, making them a modern exhibit that longtime residents still remember encountering. This exhibit, and those that followed, adopted electronics and video representation to feature some of the elements of the natural world, a trait adopted by many museums as they move through the 21st century.

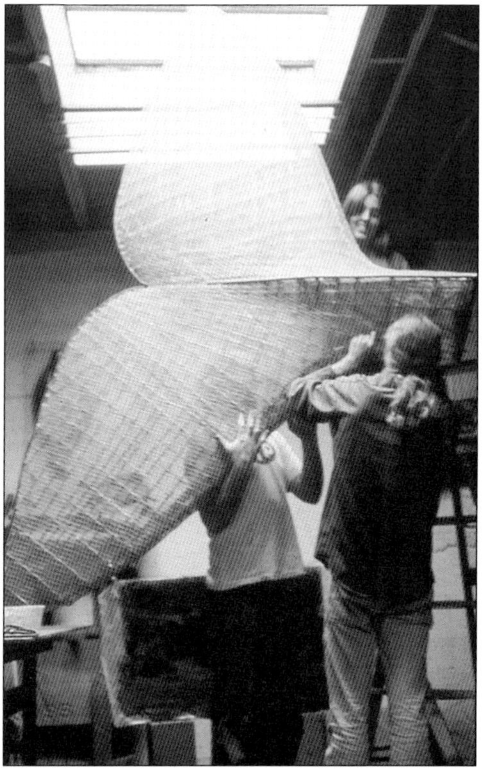

Three people work on the framework for Sandy's unfinished tail in this photograph from Foster's Oakland studio in the 1970s. After completing a whale's tail sculpture, Foster decided he would try his hand at designing and erecting an entire whale from chicken wire and ferro-cement. He completed Sandy with no contract or commission. (Courtesy Larry Foster.)

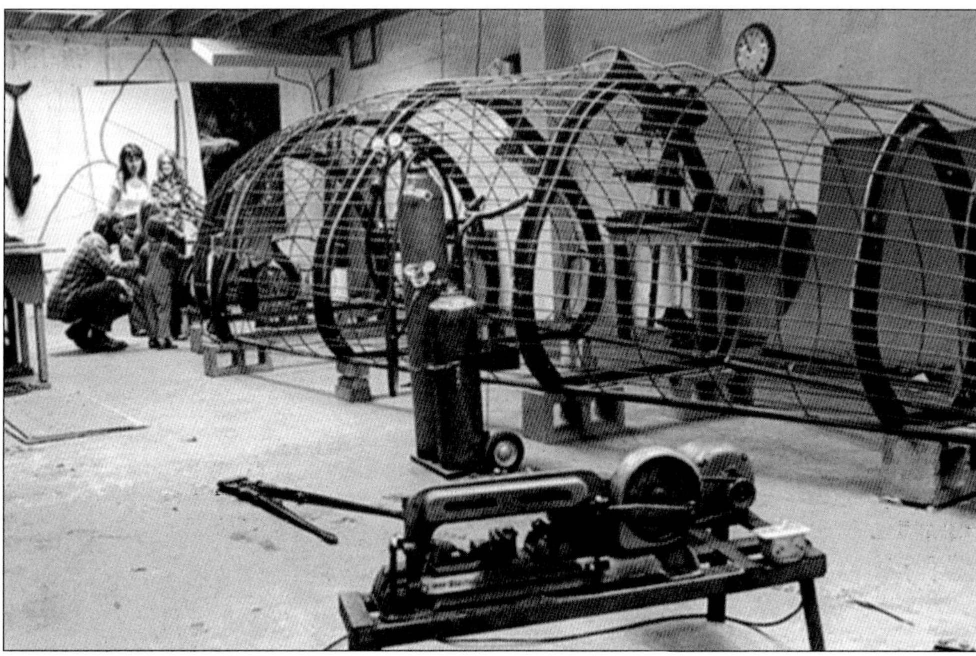

Foster's team and a young fan can be seen in this image from 1971 standing by the framework for Sandy the Whale. Foster, a former California State University professor, worked with a number of experts to develop his lifelike work on whales, including cetacean expert Ted Walker from Scripps and the museum's own curator, Vern Yadon. (Courtesy Larry Foster.)

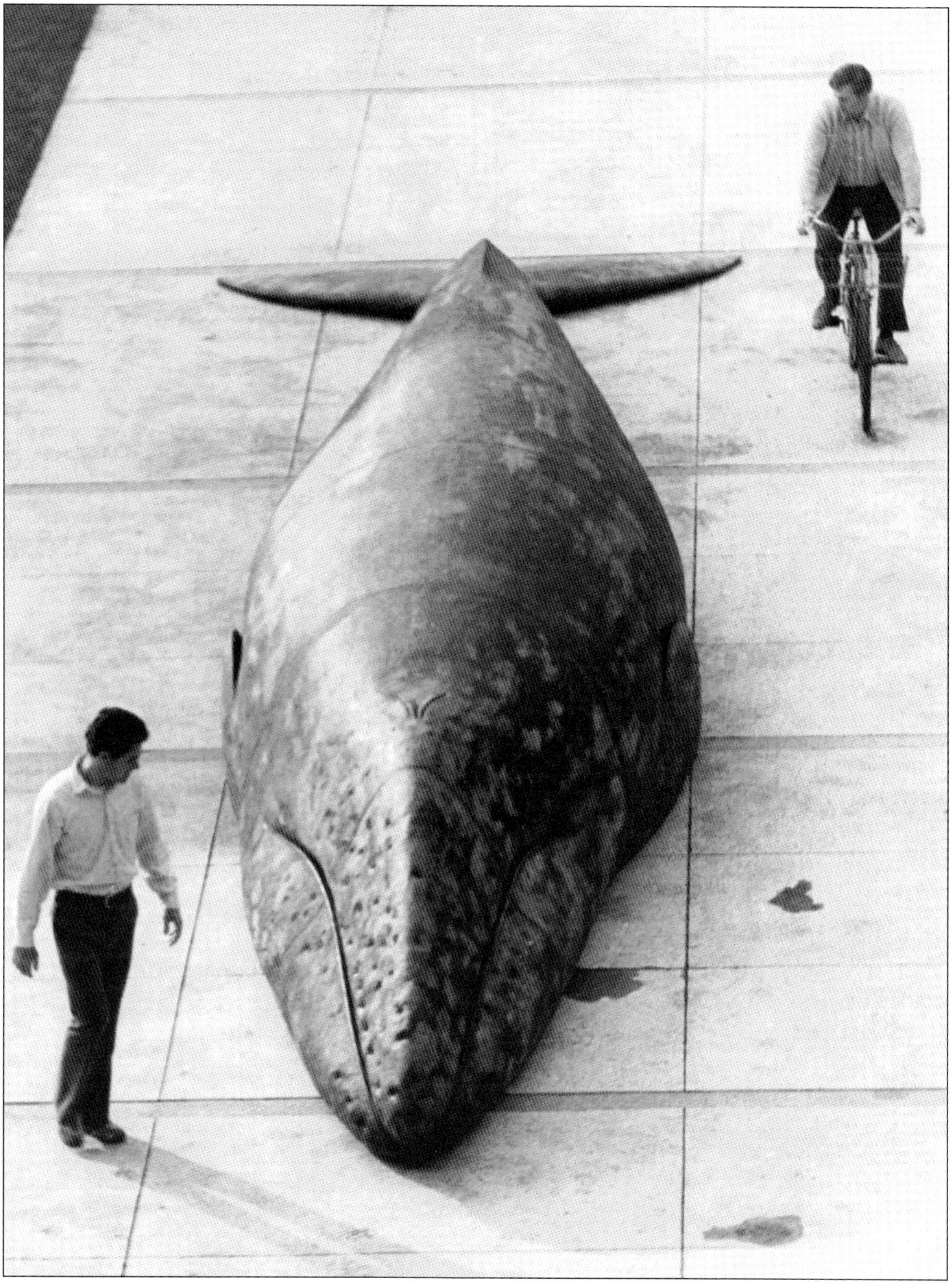

This photograph offers a glimpse into Sandy the Whale's illustrious debut at the Baxter Art Gallery, Caltech, at the Pasadena campus in 1974. Sandy offered perspective for terrestrial humans to take in the immense size and poise of the gray whale. One visitor can be seen in the photograph admiring Foster's creation, while another man on a bicycle rides past. He is also admiring the large female gray whale. As part of his General Whale company, Foster created several pieces related to the study, including a stained-glass whale, murals, educational charts, and life-sized drawings. (Courtesy Larry Foster.)

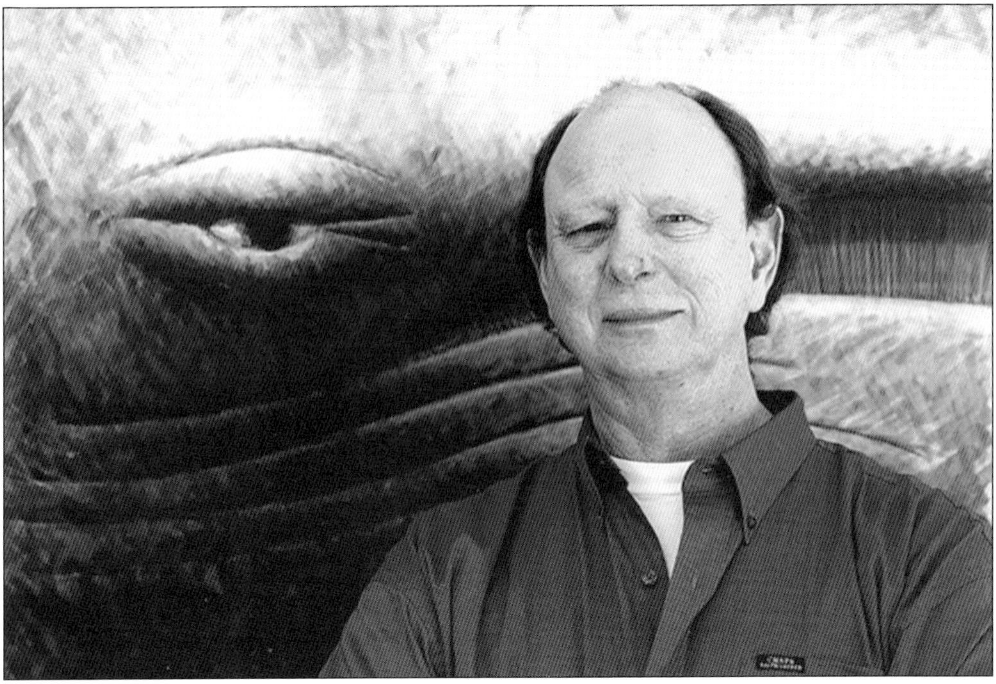

This photograph from 1999 shows artist Larry Foster with a large whale eye looking on behind him. Foster said he saw his first whale as a child. That encounter left in him a lifelong desire to learn more about the elusive sea creatures, and he wanted the world to share in his passion. (Courtesy Larry Foster.)

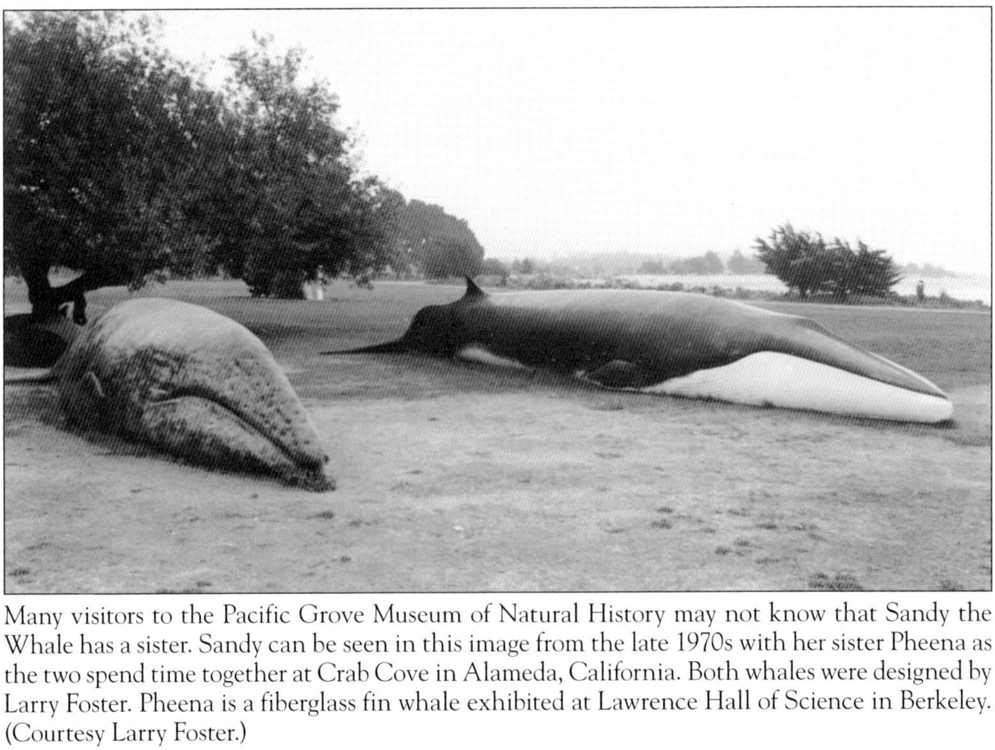

Many visitors to the Pacific Grove Museum of Natural History may not know that Sandy the Whale has a sister. Sandy can be seen in this image from the late 1970s with her sister Pheena as the two spend time together at Crab Cove in Alameda, California. Both whales were designed by Larry Foster. Pheena is a fiberglass fin whale exhibited at Lawrence Hall of Science in Berkeley. (Courtesy Larry Foster.)

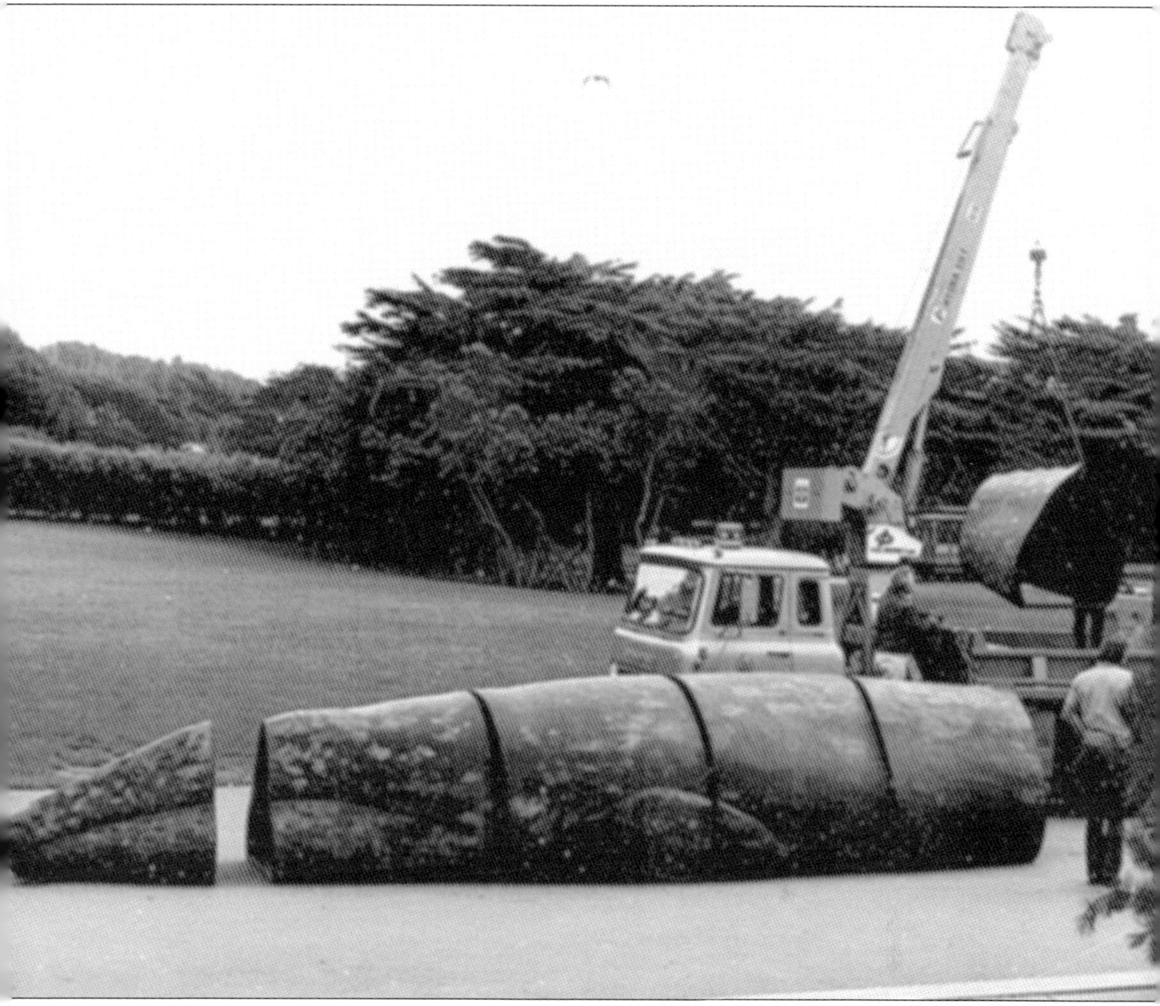

Sandy the Whale is well designed for travel, with a three-fourth-inch cement shell and a hollow core. Sandy was made from eight sections that could be unloaded from flatbed trucks. When Sandy made her debut at the California Institute of Technology, it took a truck, a forklift, and a crane to move her pieces into position for the Caltech Revisited exhibit at the university's Baxter Gallery. (Courtesy Larry Foster.)

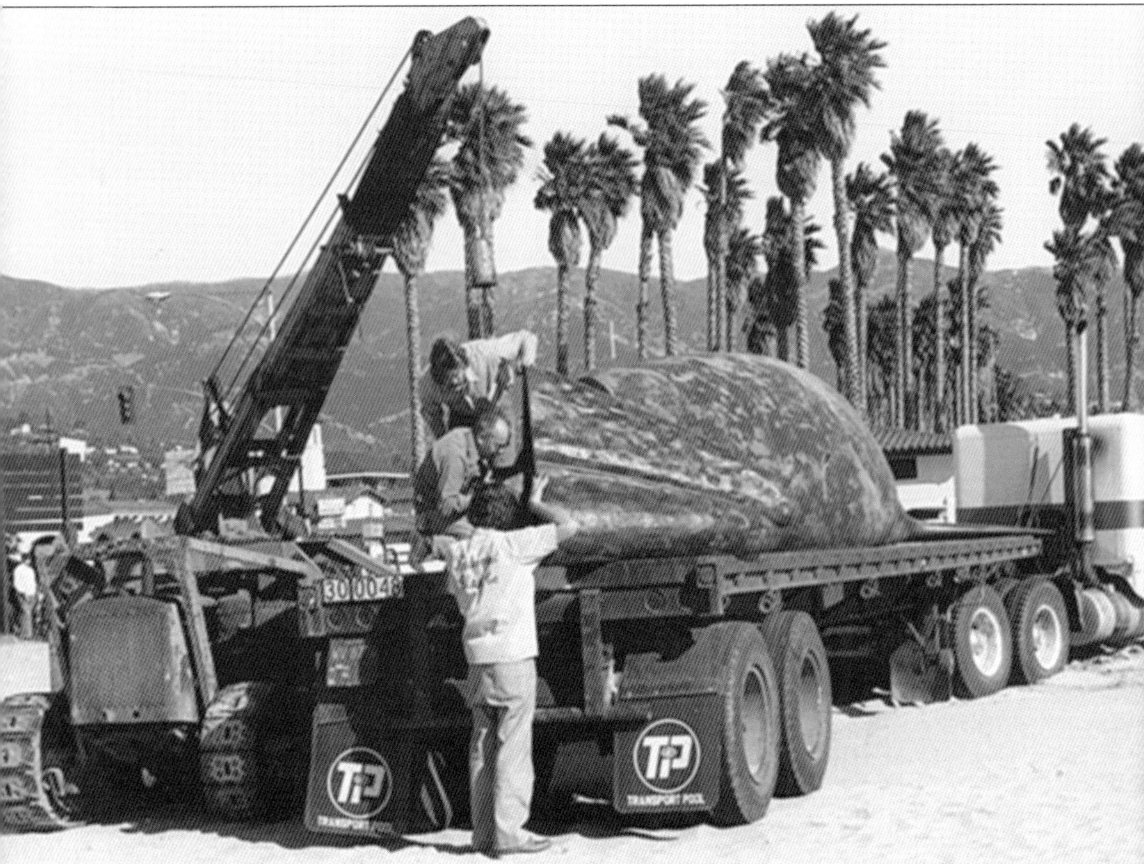

This photograph from the late 1970s shows a section of Sandy the Whale being lifted from a flatbed trailer in Santa Barbara. Her hollow head is turned in the image as she is lifted from the vehicle via a small crane to be positioned later for display. Palm trees are visible in the background. (Courtesy Larry Foster.)

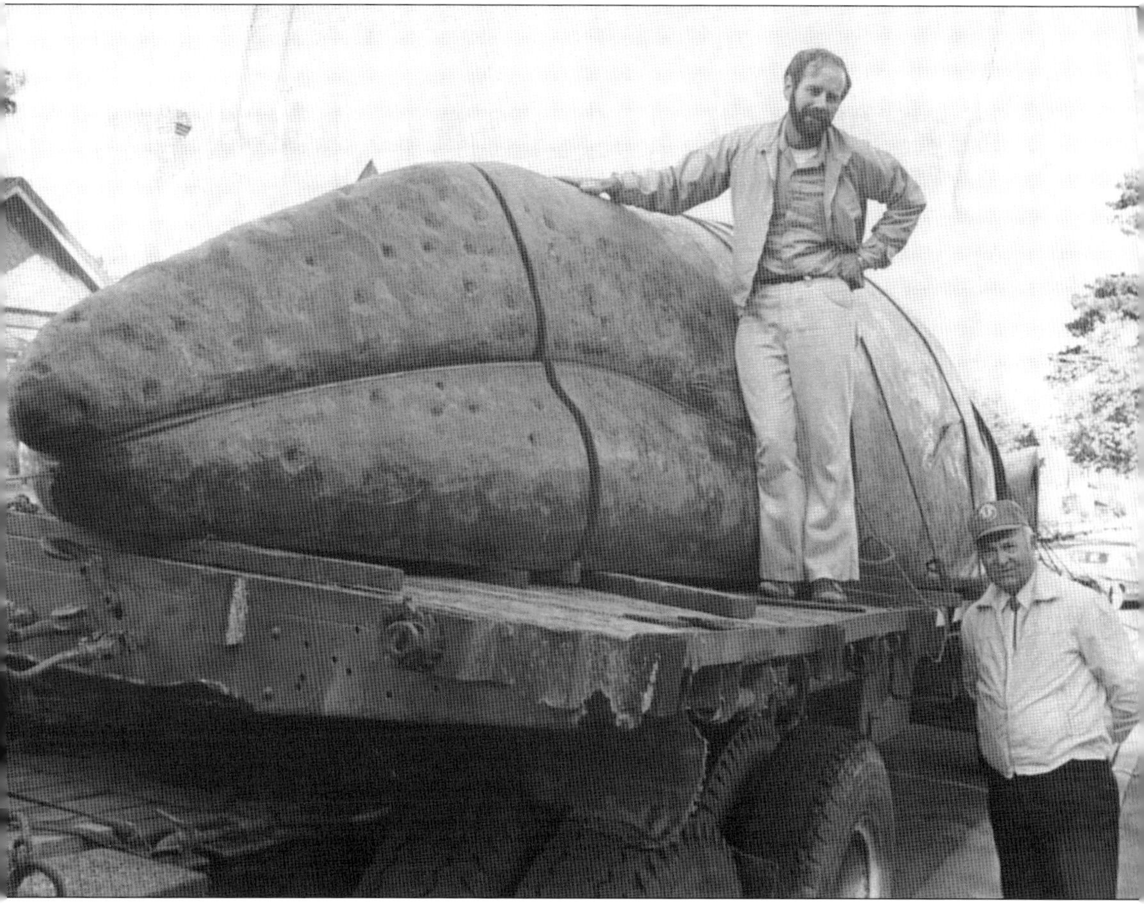

Museum curator and director Vern Yadon can be seen in this 1980s photograph that appeared in the *Monterey Herald* newspaper. Once again, Sandy is pictured resting on a flatbed trailer along Forest Avenue, where Sandy would be placed for a short period of time before she was moved to her permanent location. (Courtesy Monterey Herald.)

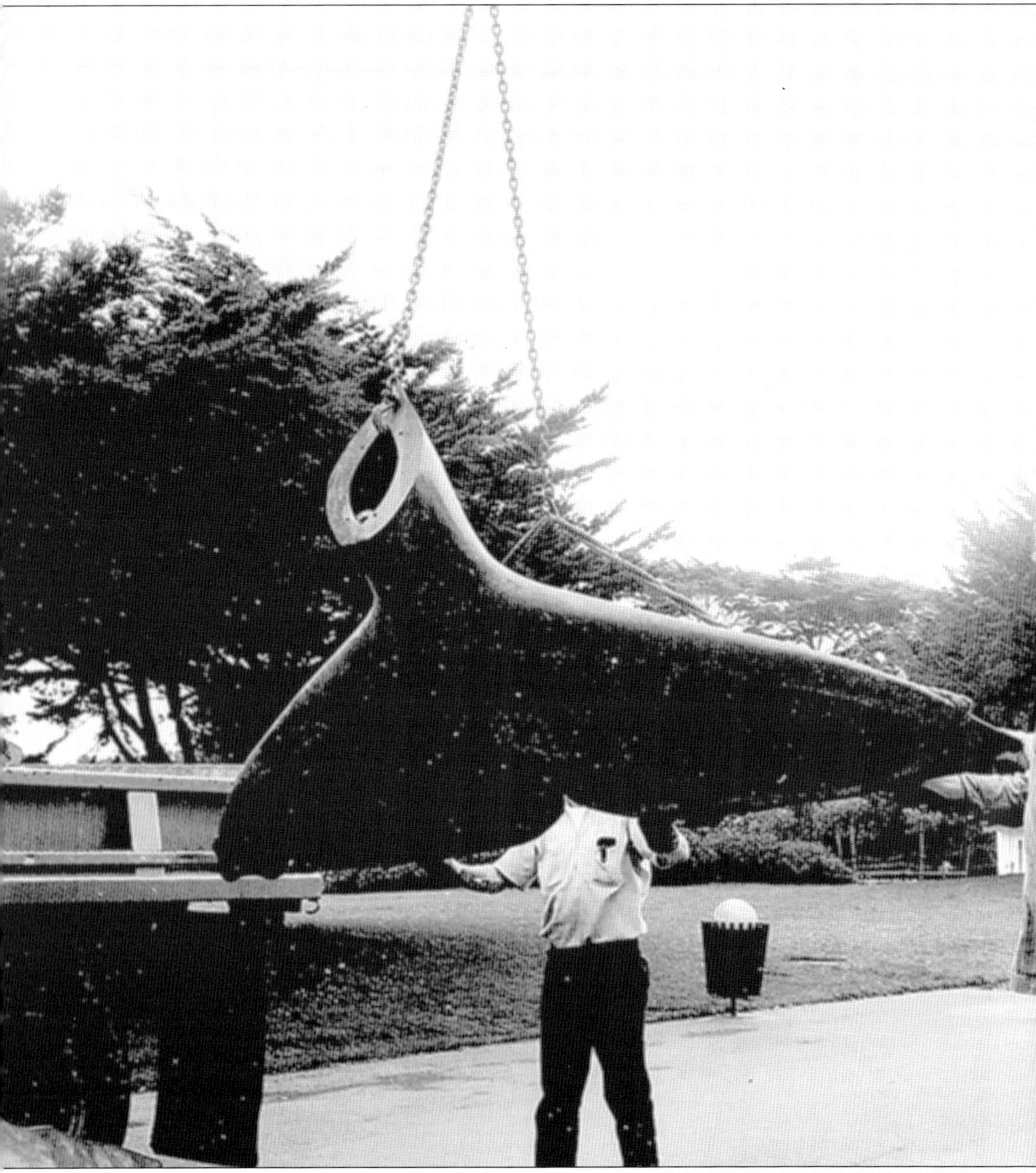

Sandy's fluke is off-loaded with the assistance of a crane in this early 1980s photograph. As part of their efforts to purchase Sandy, Pacific Grove city officials, including Mayor Florus Williams, began a campaign that invited the public to purchase a pound of the whale for $3. Those who took part were given a certificate designed by Milos Radakovich of the American Cetacean Society's Monterey Bay Chapter. (Courtesy Larry Foster.)

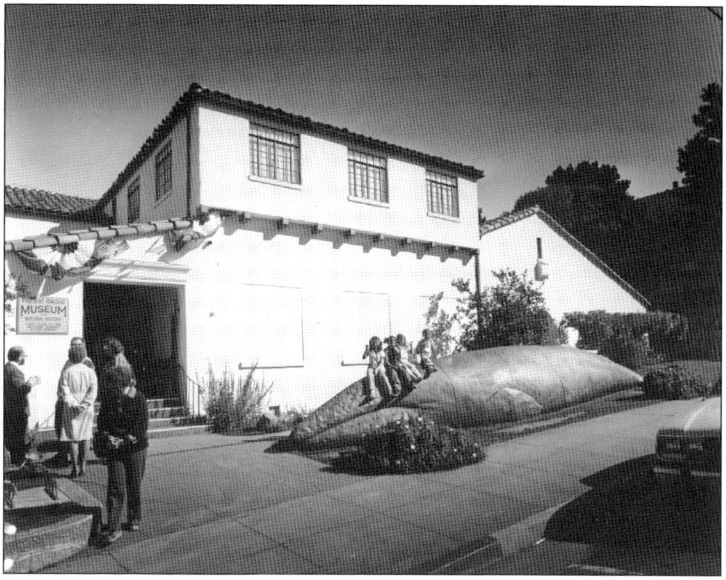

This photograph by Jerry Lebeck shows Sandy the Whale in the early 1980s when the concrete sculpture was located on Forest Avenue. The old entrance to the museum can viewed on the left. Sandy was later moved to the front of the museum along Central Avenue, where she remains. (Courtesy Pacific Grove Museum of Natural History.)

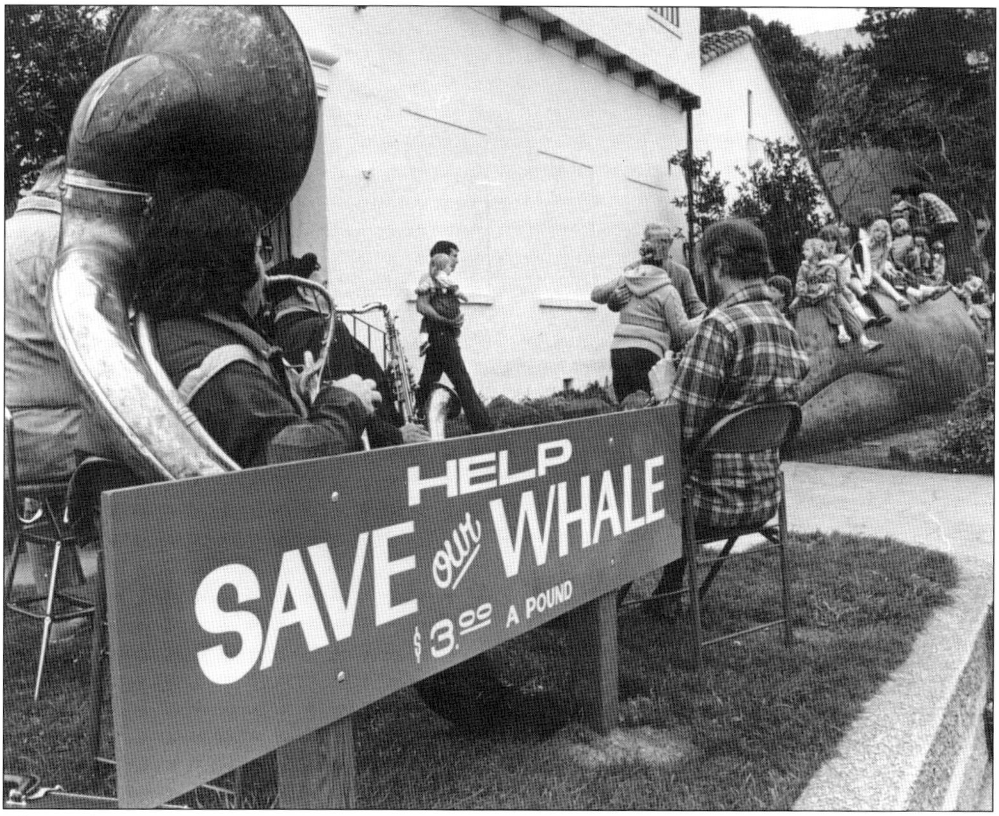

This early-1980s photograph from the *Monterey Herald* newspaper helped spread the word about a museum fundraiser to purchase Sandy the Whale and give the sculpture a permanent home in Pacific Grove. The words "Help Save our Whale" can be seen on a sign in the photograph. Community members could buy "a pound" of Sandy during the fundraiser and were given a certificate for their donation. (Courtesy the *Monterey Herald*.)

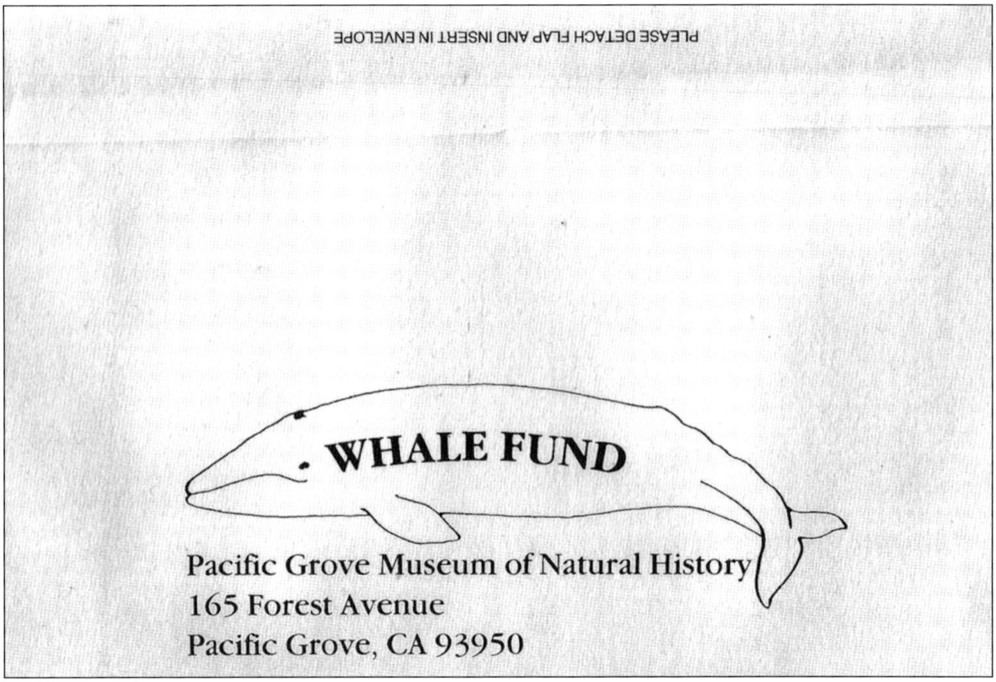

Pictured in this image from the early 1980s is a Whale Fund card that allowed the friend of the Pacific Grove Museum of Natural History to purchase a pound of Sandy the Whale. To this day, many Pacific Grove residents fondly remember the effort to buy the large whale for permanent display and still own the certificate they received for contributing. (Courtesy Pacific Grove Museum of Natural History.)

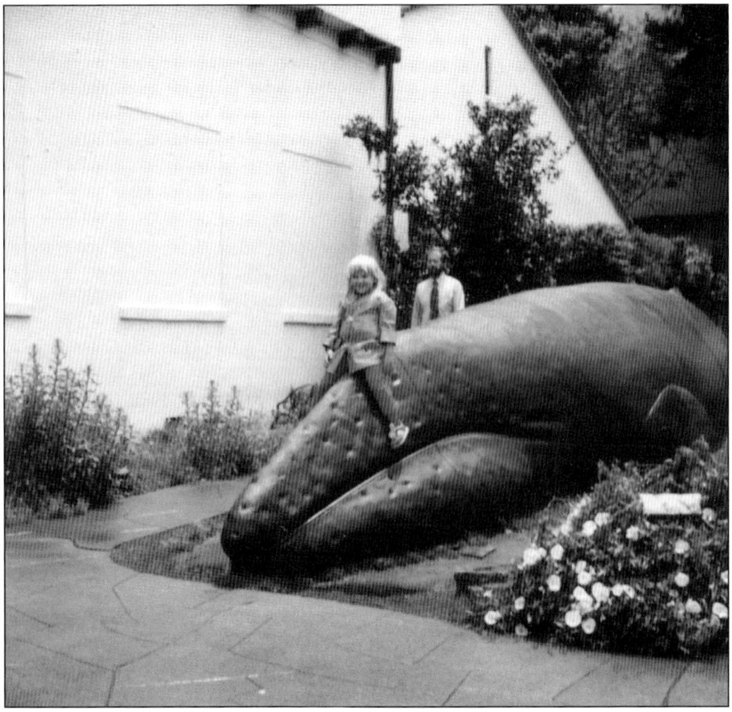

This summer 1982 photograph shows Sandy the Whale with a young girl, A. Norton, riding atop her head. Curator and director Vern Yadon can be seen standing behind the whale. Wildflowers are visible to the right of the photograph. This picture was taken before Sandy moved to her Central Avenue location in 1983. (Courtesy Pacific Grove Museum of Natural History.)

While Sandy the Whale originally started her life at the museum along Forest Avenue, she did not stay there for long. Sandy was moved to her permanent location along Central Avenue in 1983. This 1983 image shows a view of the drainpipe constructed around the large gray whale. (Courtesy Pacific Grove Museum of Natural History.)

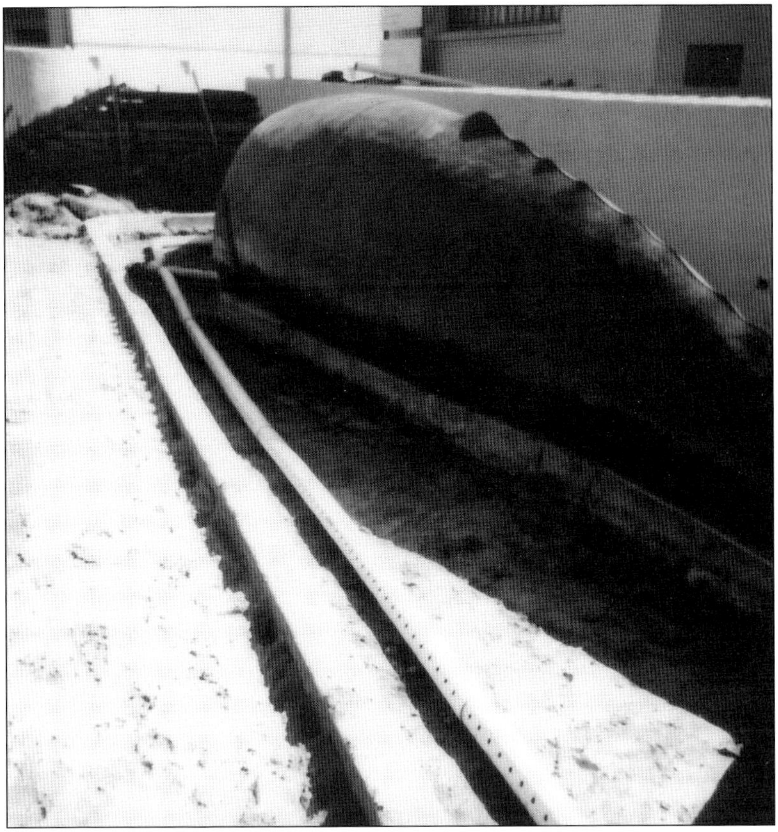

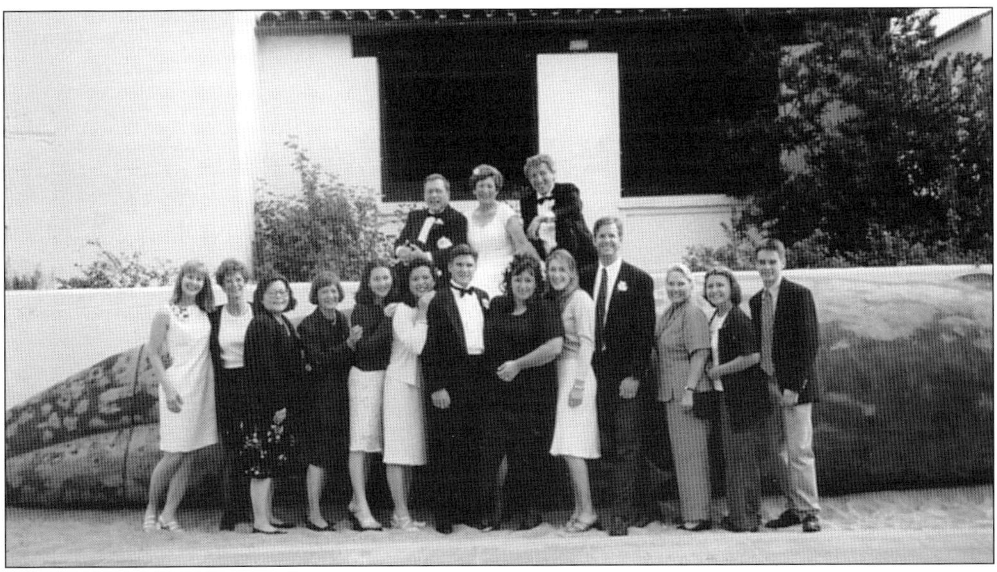

Pictured is the marriage of Maria and Frank Penner in front of Sandy the Whale. The two were married on July 30, 1999, with a reception at the Old Bath House, which is today known as the Beach House. Frank and Maria Penner are visible sitting atop Sandy surrounded by family members. The museum continues to offer weddings on site for interested parties to this day. (Courtesy Frank and Maria Penner.)

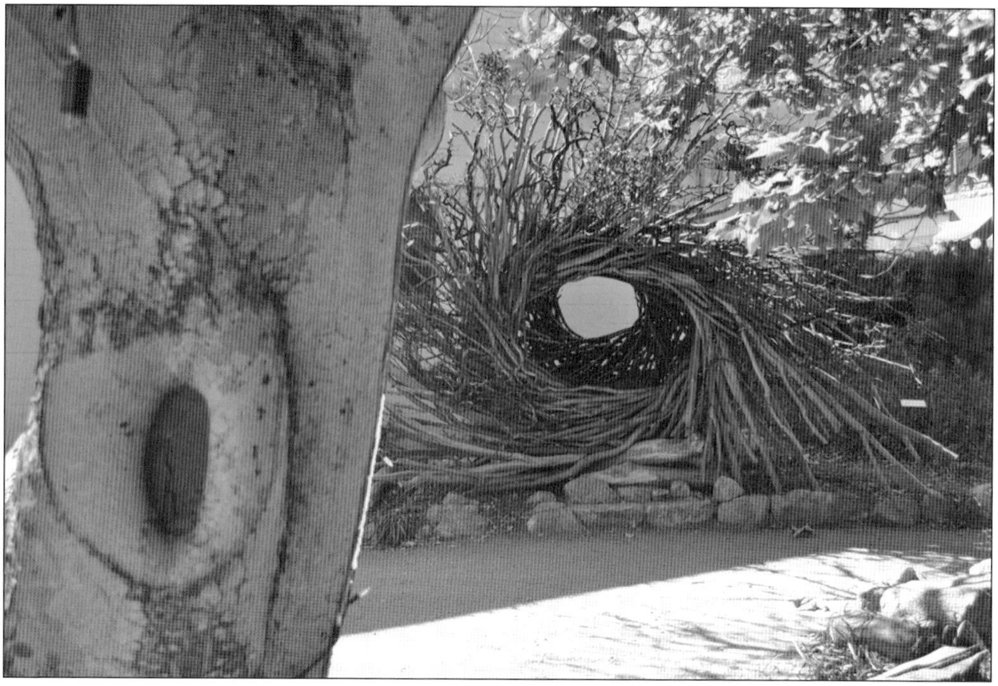

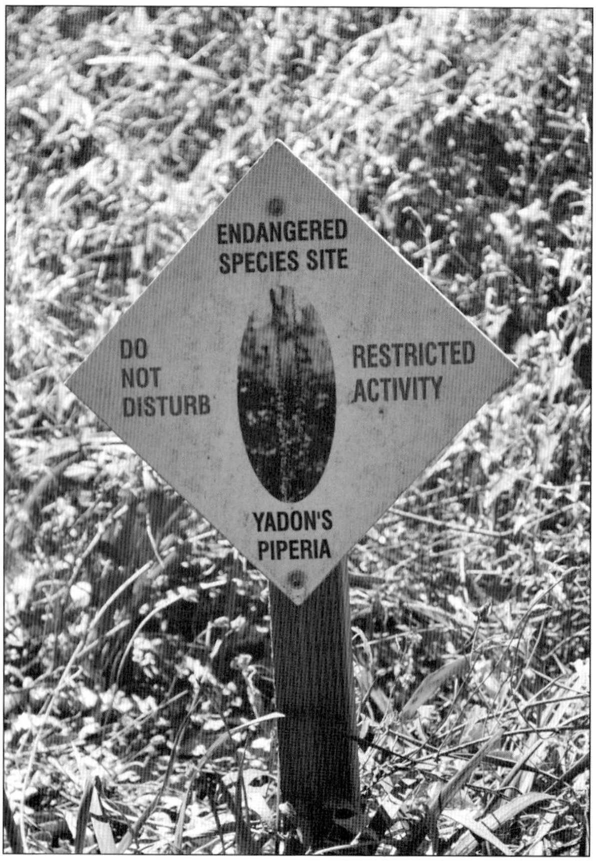

The Native Plant Garden has been a part of the natural history museum since the late 1800s. The garden received a sizable upgrade in the 1950s that included hundreds of native plants found in Monterey County. Curator and director Vern Yadon is credited with overhauling the Native Plant Garden at the time. The large buckeye tree, visible in the foreground, reportedly came with the original lot. In the background of the image is the Spirit Nest, made by local artist Jason Fann. (Courtesy Pacific Grove Museum of Natural History.)

Vern's Piperia is also planted at the museum in the Native Plant Garden. Named for director and curator Vern Yadon, the area where the piperia is planted is restricted from public activity. There are two locations in the garden where the piperia can be found in potted containers. The piperia is clearly marked as an endangered species. (Courtesy Pacific Grove Museum of Natural History.)

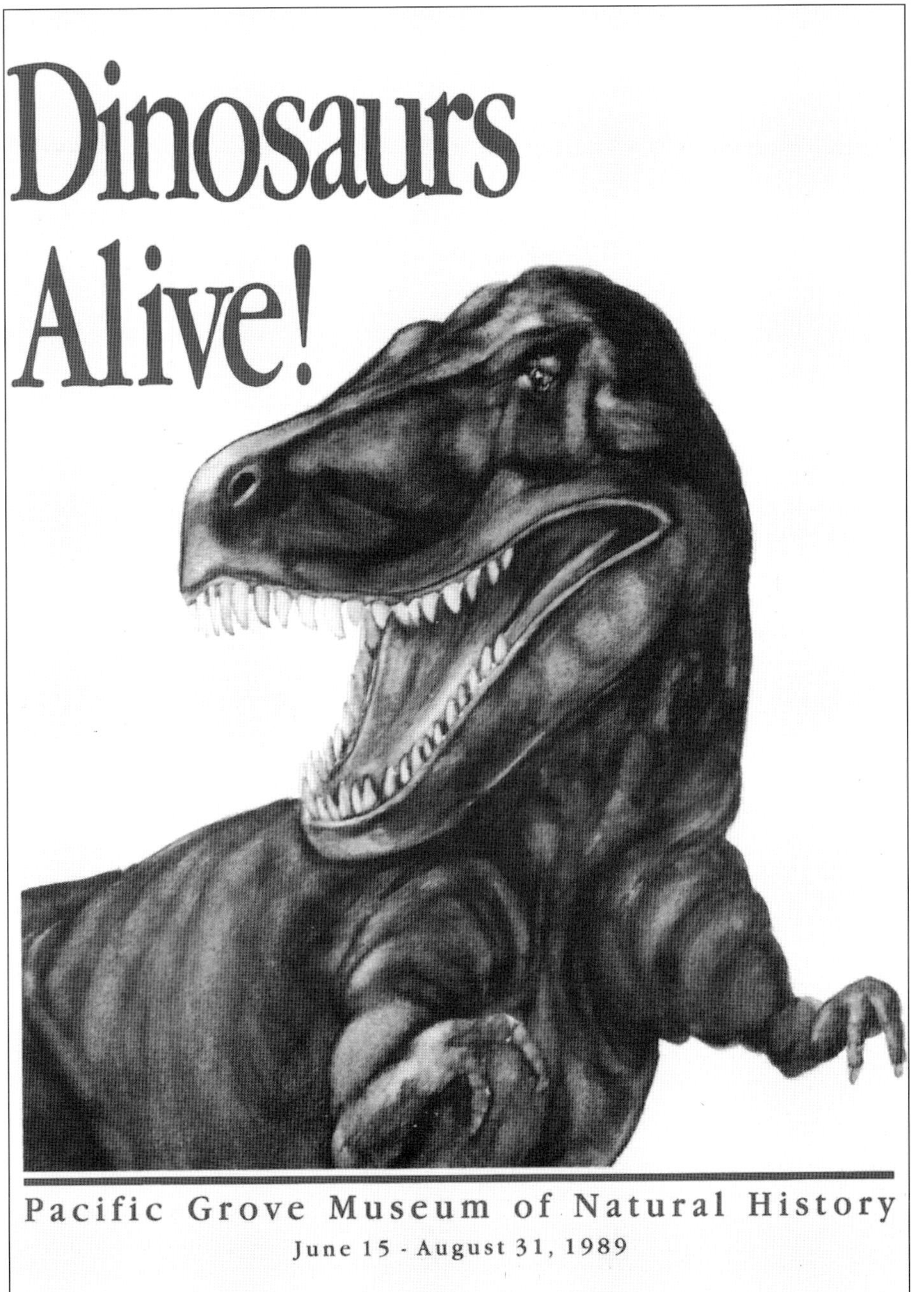

In the 1980s, the museum played host to other large exhibits besides Sandy the Whale. From June through August 1989, museum guests were treated to the fantastic Dinosaurs Alive! exhibit. A special trifold informational card was created to attract visitors to the event, and thousands would turn out during the special engagement. The card featured brief introductory information related to the group of dinosaurs that stomped into the museum and focused on the species visitors could expect to see. Ticketron tickets were made available to guests prior to visiting the museum to expedite their adventure to Pacific Grove. (Courtesy Pacific Grove Museum of Natural History.)

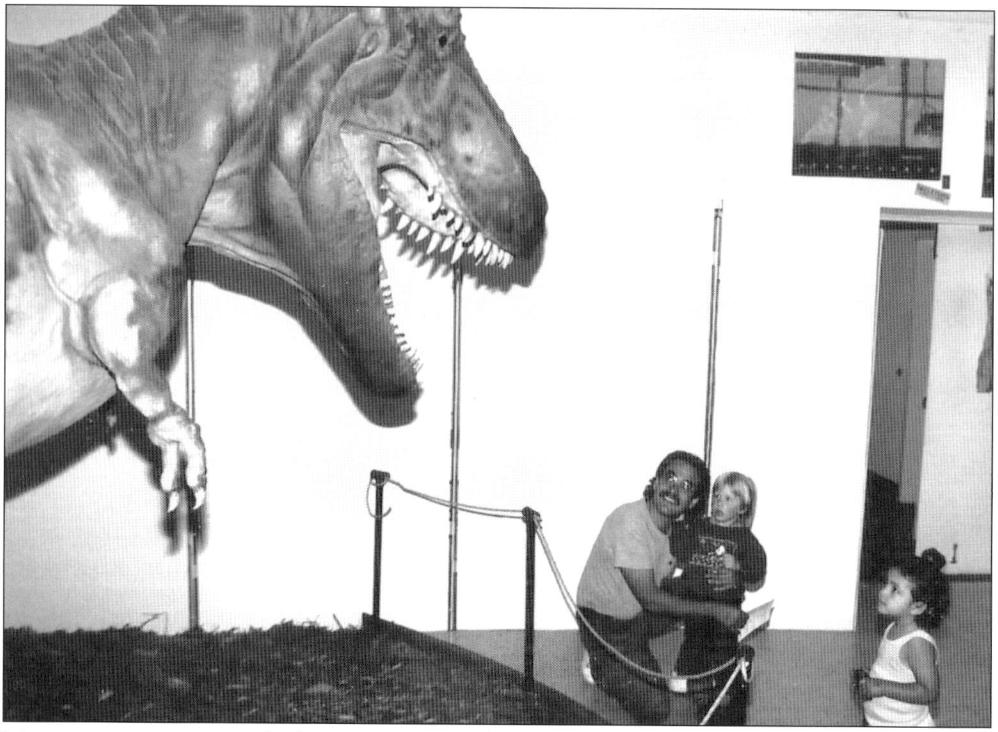

Museum visitors, some a little unsure, meet a large prehistoric friend in this 1989 photograph. Then museum curator Paul Finnegan worked for five years to bring the mechanized dinosaurs to the museum. The large exhibit brought thousands of visitors into the museum from far and wide. (Courtesy Pacific Grove Museum of Natural History.)

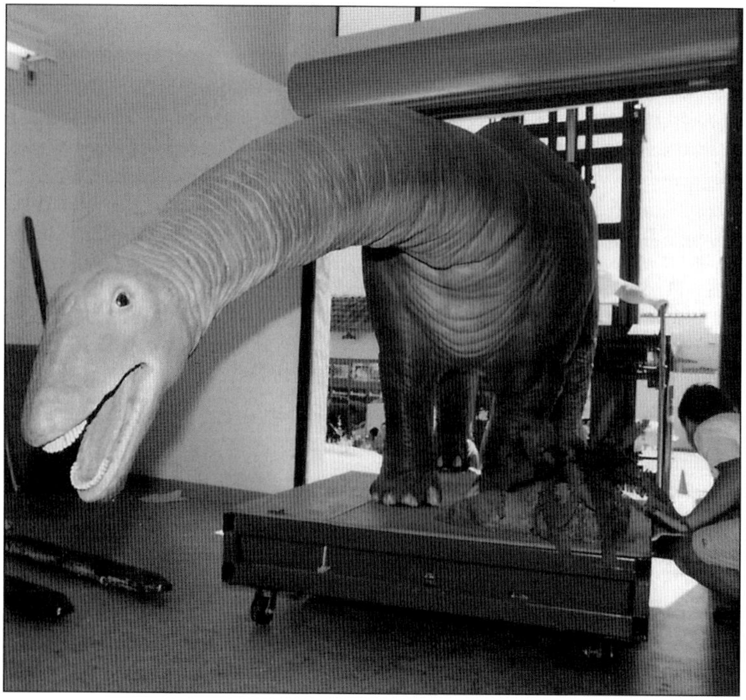

This 1989 photograph illustrates some of the difficulty involved in bringing these Mesozoic creatures inside the natural history museum. Rolling doors located on the Grand Avenue side of the museum were helpful in allowing entry to this brontosaurus. The tyrannosaurus was too large to fit in the gallery space provided to it, so he was placed in the lobby. (Courtesy Pacific Grove Museum of Natural History.)

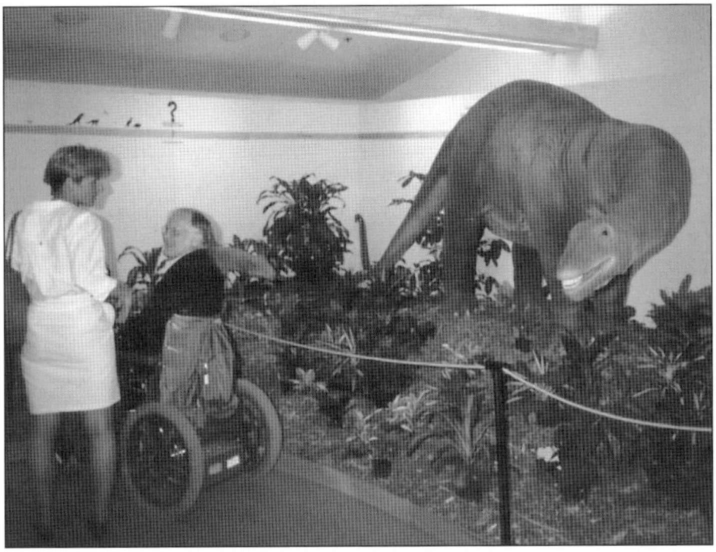

The 1989 exhibit was such a hit with museum guests that adults also found themselves swept up in the fun. According to accounts, dinosaur parties were held during the Dinosaurs Alive! exhibit. These parties included dinosaur costumes and "dino-food." The museum store also turned into a "dino-shop" during the exhibit. (Courtesy Pacific Grove Museum of Natural History.)

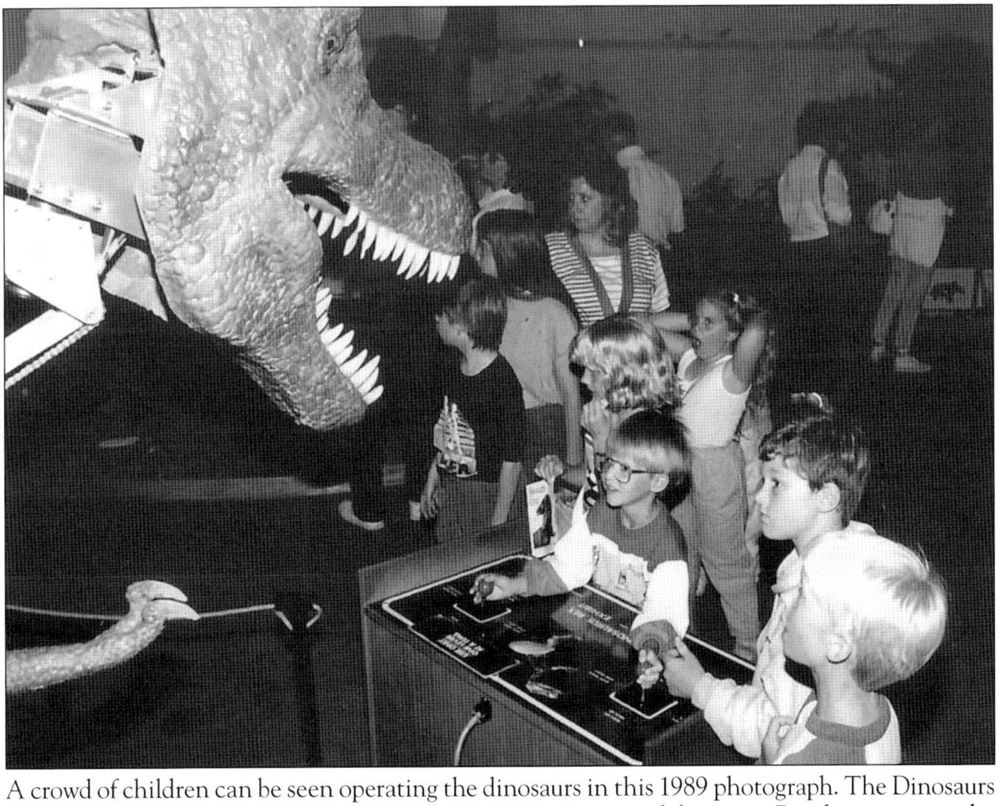

A crowd of children can be seen operating the dinosaurs in this 1989 photograph. The Dinosaurs Alive! exhibit remained at the museum from June to August of that year. Besides operating the dinosaurs, children were also invited to take part in activities that granted them a dinosaur "hunting license," which allowed them to identify the creatures they saw in the museum. (Courtesy Pacific Grove Museum of Natural History.)

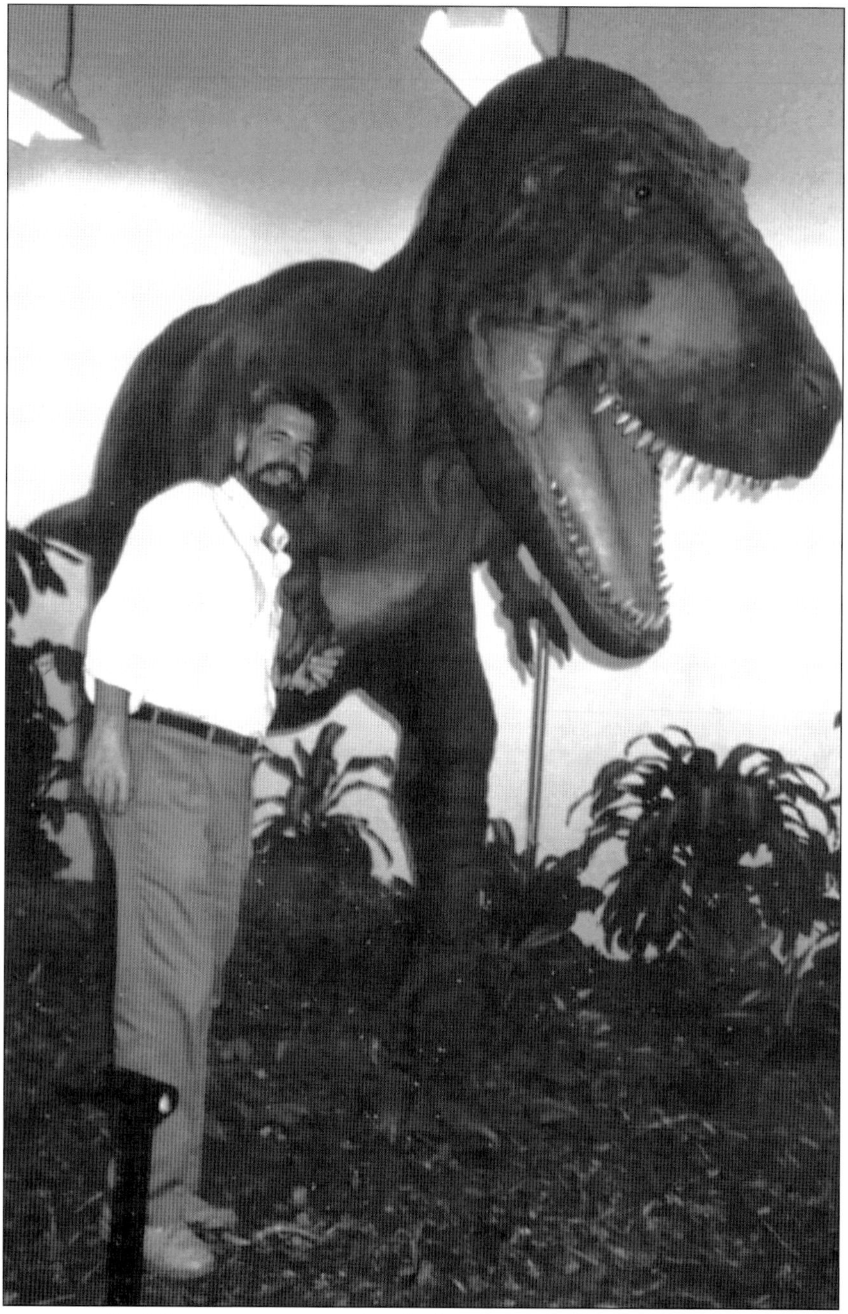

Museum curator Paul Finnegan poses with a hulking tyrannosaurus, otherwise known as the "Tyrant Lizard," in this 1989 photograph. The large dinosaurs on exhibit were considered state of the art at the time and constructed to look, move, and even feel like the real thing based on scientific data. Besides the prehistoric dinosaurs themselves, "dinamation" technicians also traveled with the Dinosaurs Alive! exhibit to ensure the creatures stayed in control. The large-scale museum event was made possible thanks to the museum and many other supporters, including the Pebble Beach Foundation, the Pacific Grove Chamber of Commerce, and others. (Courtesy Pacific Grove Museum of Natural History.)

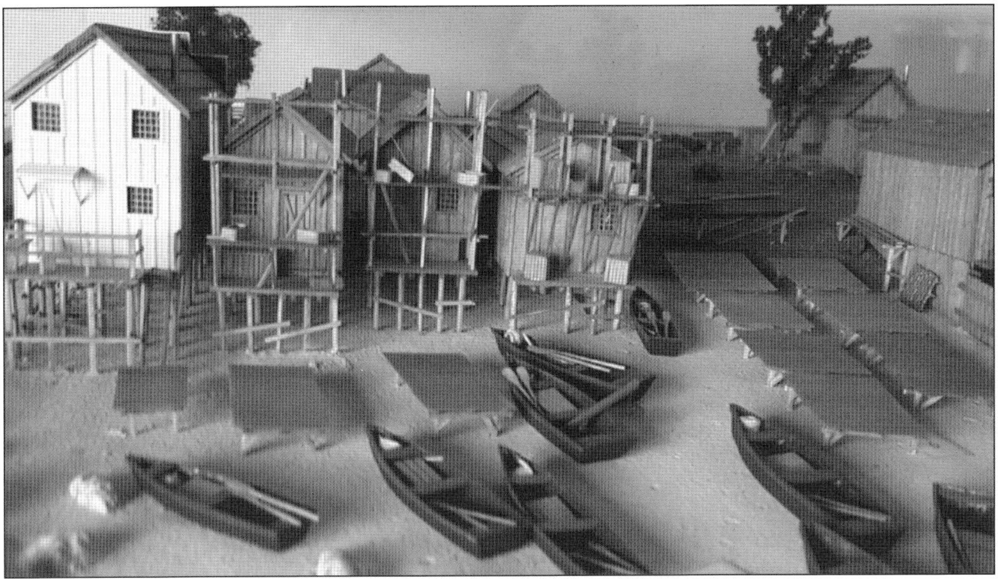

Pictured is the model of the Chinese Fishing Village made by artist Michael E. Croft. It has been on display since 2012, according to museum records. The village located at Point Alones was considered a unique one among early Chinese settlements in that, not just bachelors, but also children and women lived in the village. (Courtesy Pacific Grove Museum of Natural History.)

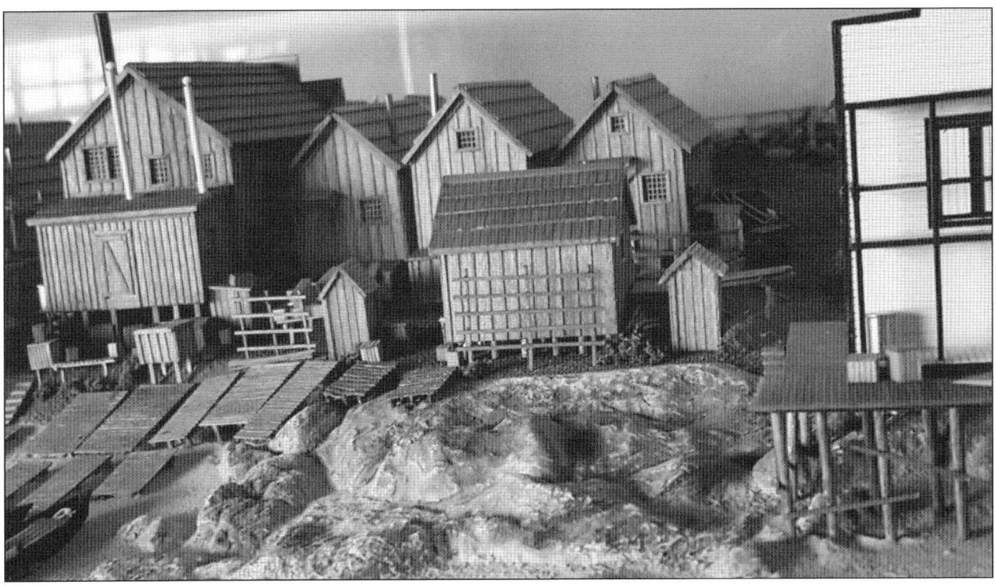

Croft's 2012 model of the Chinese Fishing Village is a favorite among museum visitors and is part of an exhibit dedicated to the early Chinese inhabitants of Pacific Grove. Every year, the museum holds the "Walk of Remembrance" with Gerry Low-Sabado, a Chinese Fishing Village descendant. The walk begins at the museum and travels to the former site of the village along the shoreline. (Courtesy Pacific Grove Museum of Natural History.)

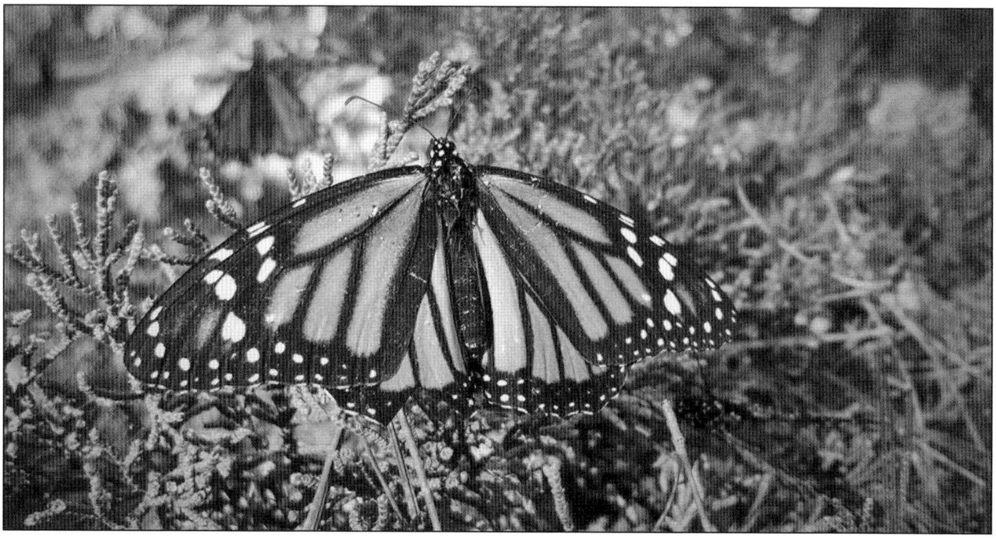

In 2008, the museum began to staff the Monarch Grove Sanctuary, found at 250 Ridge Road off Lighthouse Avenue, with trained docents. Prior to that, the organization Friends of the Monarchs staffed the popular wooded area during the butterfly overwintering season. These volunteers can be found assisting visitors at the sanctuary most days, from around noon until 4:00 p.m., from November through February. (Courtesy Pacific Grove Museum of Natural History.)

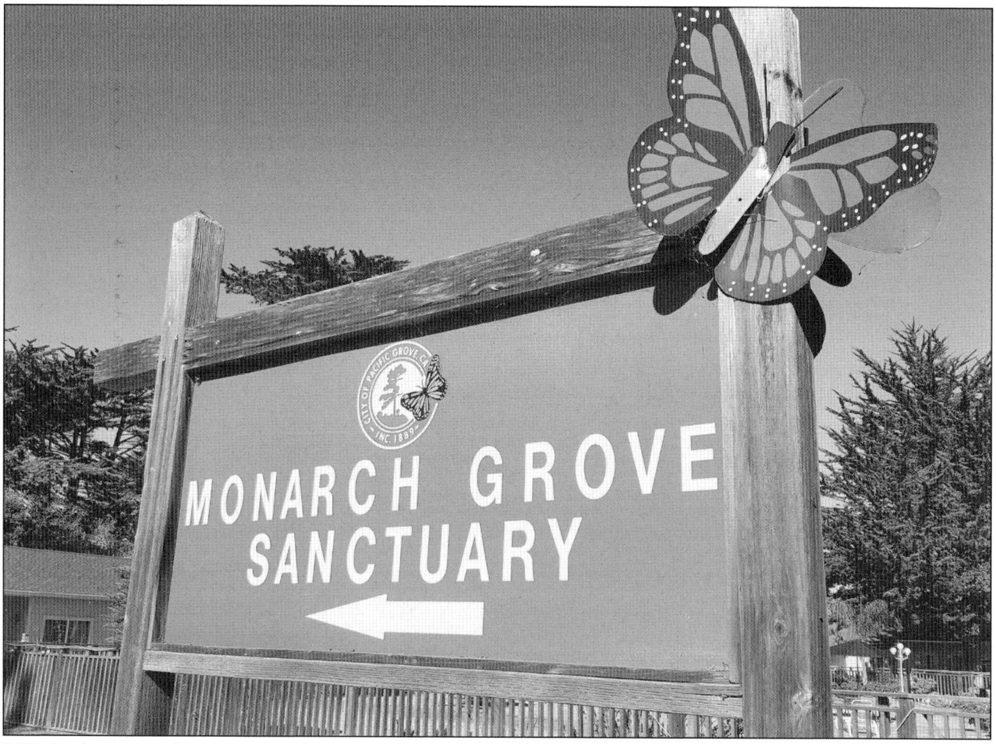

While their numbers vary each year, the number of monarchs has diminished over the years. The City of Pacific Grove deeded an easement for the monarch butterflies in 1992 that prevented development on the site. Thousands of visitors make their way to the sanctuary every year to view the monarchs. (Courtesy Pacific Grove Museum of Natural History.)

In 1997, another new exhibit featured items and educational materials related to the whales found in Monterey Bay. The Whale Room featured whale sounds, tiled whale art, gull taxidermy, and a nearly complete skeleton of a killer whale. The room was eventually turned into the museum store, while many of the items have been moved to other locations and returned to the basement collection. (Courtesy Pacific Grove Museum of Natural History.)

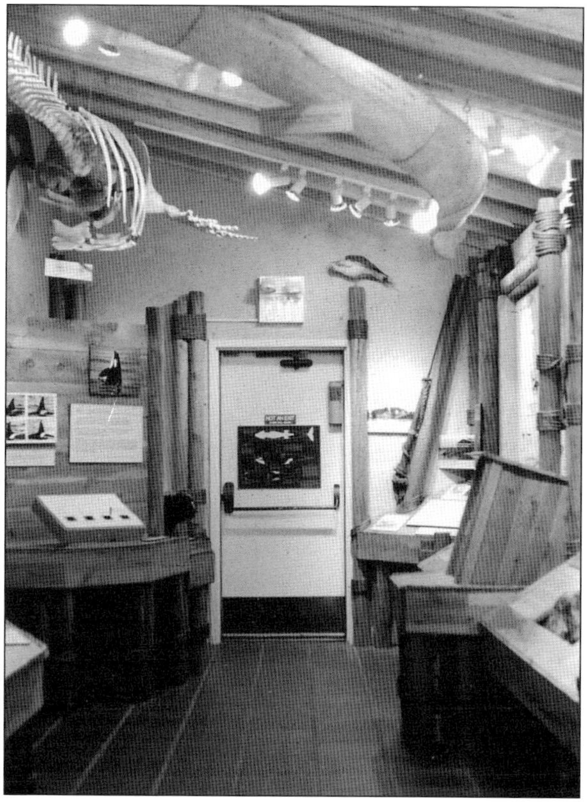

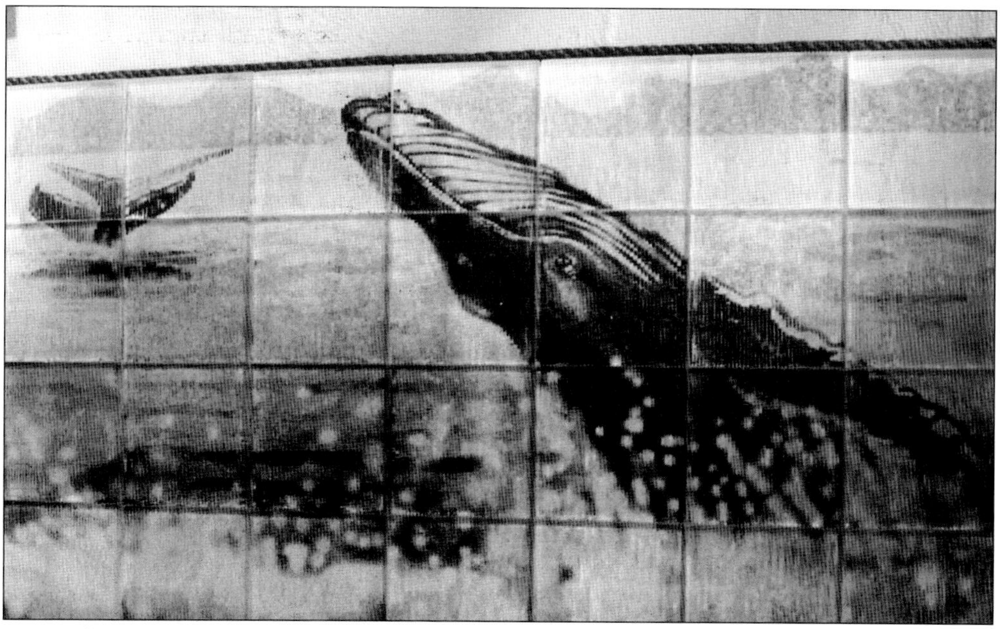

Pictured are "life tiles" that appear to move when seen from different angles. The items were installed in 1997 and remain in the museum's collection, though they are no longer on display. Artist Rufus Butler Seder created the humpback whale ("life tile") mural using 32 wall-mounted, laser-impregnated glass tiles. (Courtesy Pacific Grove Museum of Natural History.)

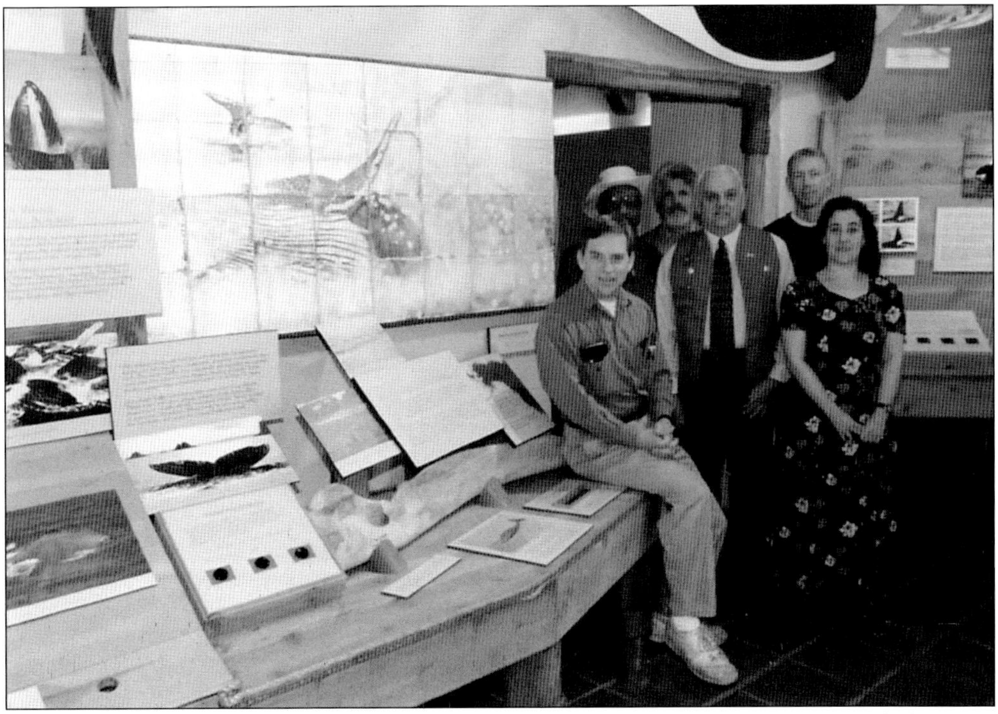

Museum staff, including Dr. Stephen Bailey, curator, can be seen enjoying the new Whale Room in 1997. Bailey is pictured seated next to the life tile mural. Above the group is a skeleton of a killer whale, which remains in the room today. Behind the group are buttons that allowed visitors to hear the sound of whales when they pressed them. (Courtesy Pacific Grove Museum of Natural History.)

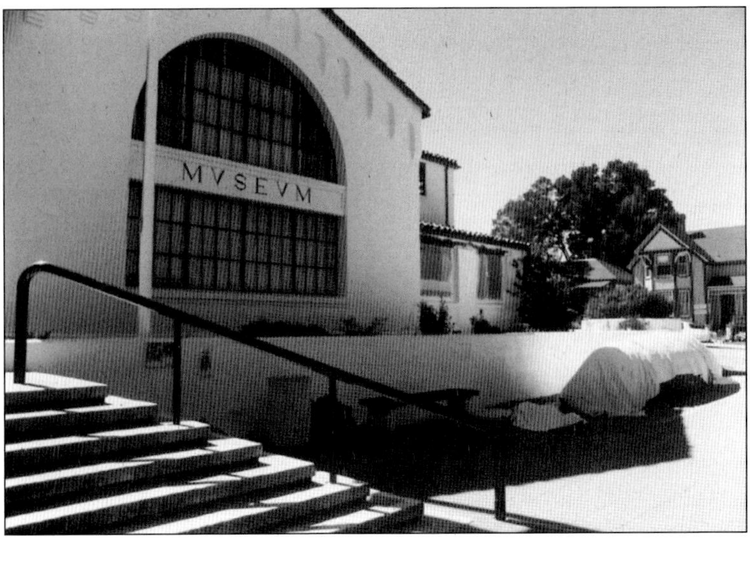

Sandy the Whale is pictured under a painter's tarp in 1997. The museum got a fresh paint job that year that changed the building's exterior color from white to beige, matching the Pacific Grove Public Library across the street. (Courtesy Pacific Grove Museum of Natural History.)

Museum staff welcomed a flag-raising ceremony in front of the museum, near the stairs leading to the main entrance, in 1995. According to museum records, pictured are Ramon Ayres, Easter Brock, Demele, and Larson. The flagstaff plaque shows that is in memory of Gen. Joseph W. Stilwell of the US Army. (Courtesy Pacific Grove Museum of Natural History.)

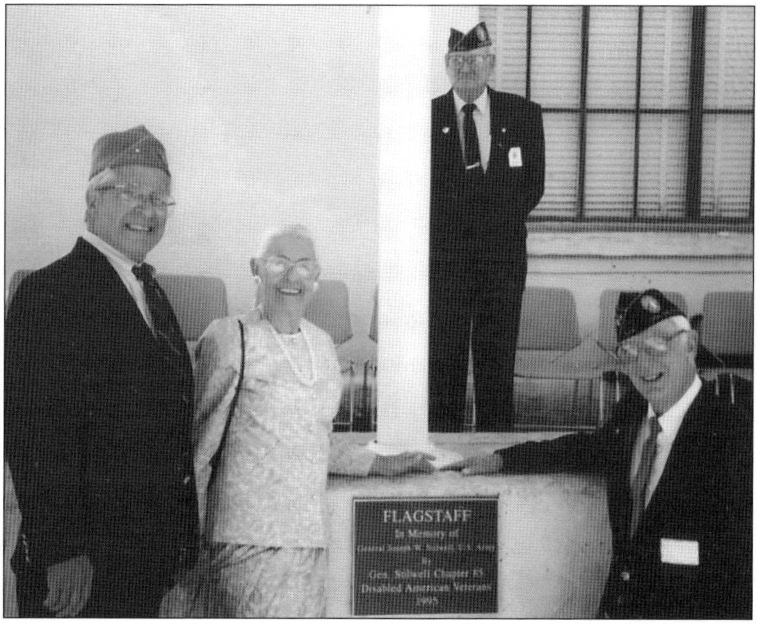

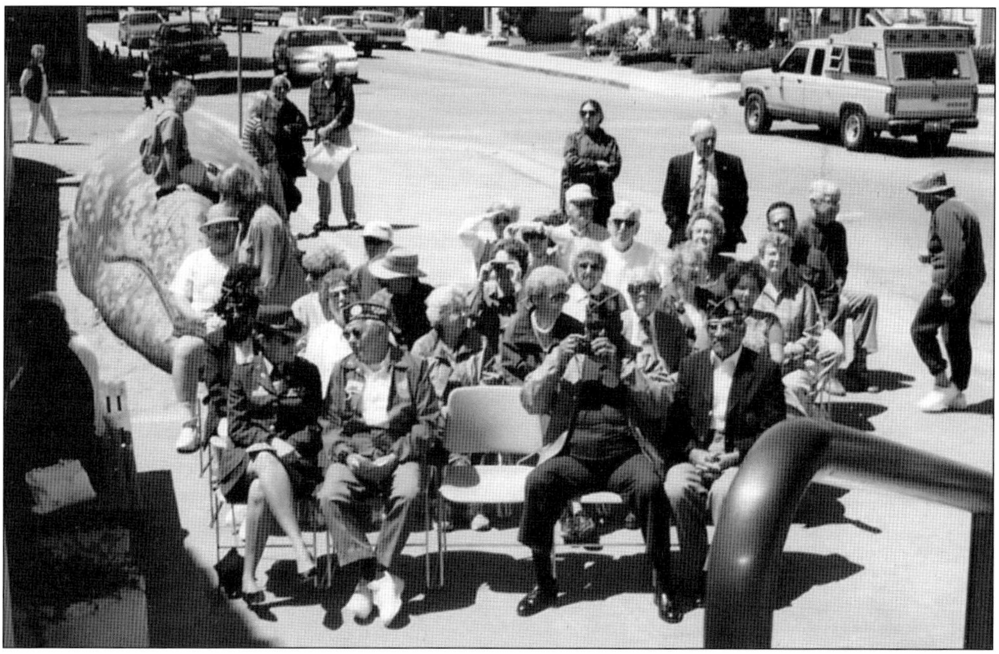

A crowd gathers around Sandy the Whale in this photograph of the 1995 flagstaff installation. Those in attendance included Moe Ammar with the Pacific Grove Chamber and Col. Ila Mette, garrison commander for Fort Ord and the Presidio at the time. The chamber of commerce's current location can be seen in the background. (Courtesy Pacific Grove Museum of Natural History.)

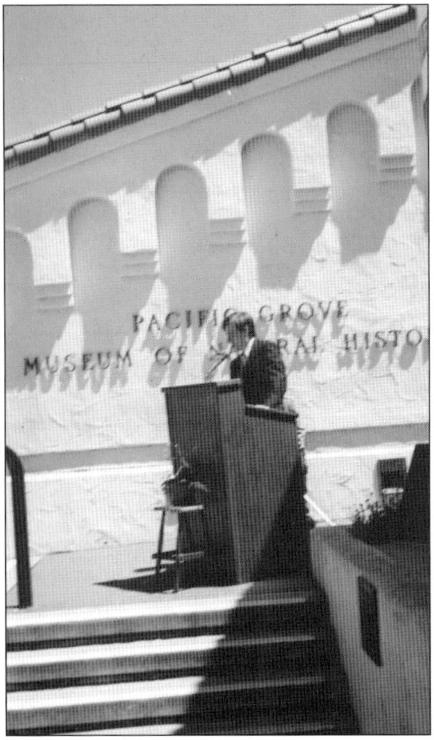

Curator Stephen Bailey speaks to the crowd during the August 1995 Flagstaff at the front of the museum. The flagstaff was made possible by the Gen. Stilwell Chapter 85 of Disabled American Veterans. The plaque and flag can still be seen flying over the museum to this day. (Courtesy Pacific Grove Museum of Natural History.)

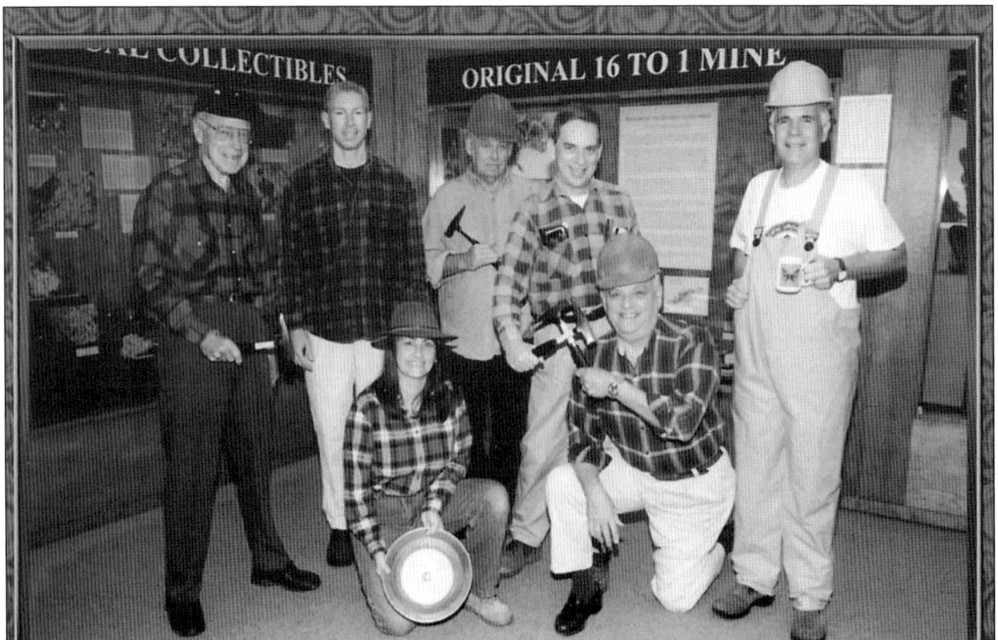

Museum staff can be seen hamming it up for the camera in this holiday greetings photograph from the early 1990s. The group is dressed as if they are ready to mine for gold in the image. They are pictured standing before a glass-encased exhibit featuring information on mining and geology. (Courtesy Pacific Grove Museum of Natural History.)

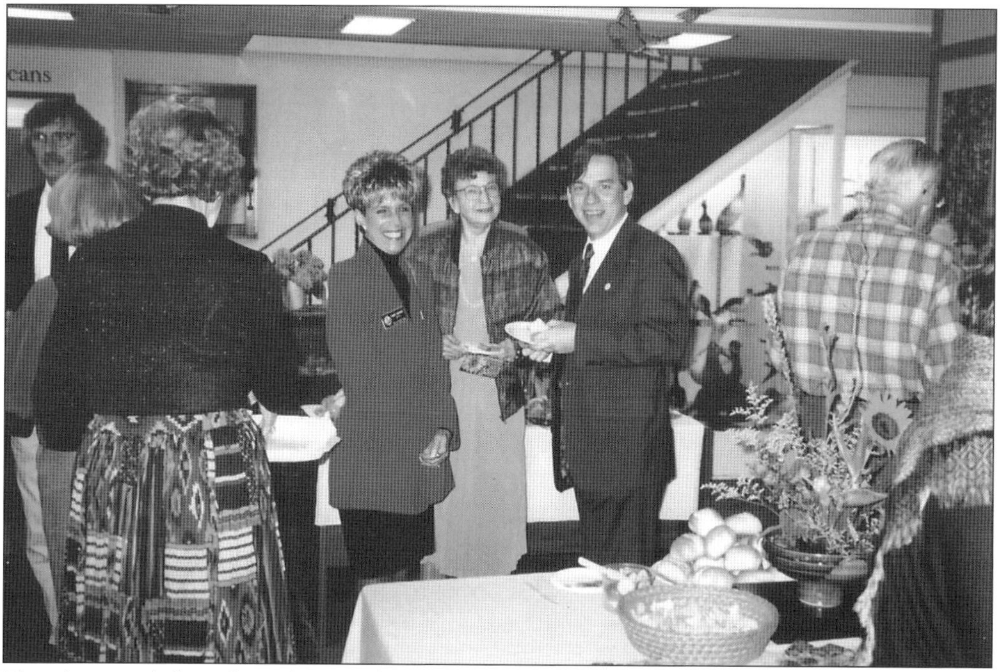

This photograph from the early 1990s shows a special event, possibly an exhibit opening, from the early 1990s. Curator Stephen Bailey can be seen entertaining two guests at the center of the photograph. A spread of food is also visible sitting at the center of the Heritage Gallery at the entrance to what was then called the Room of Man. (Courtesy Pacific Grove Museum of Natural History.)

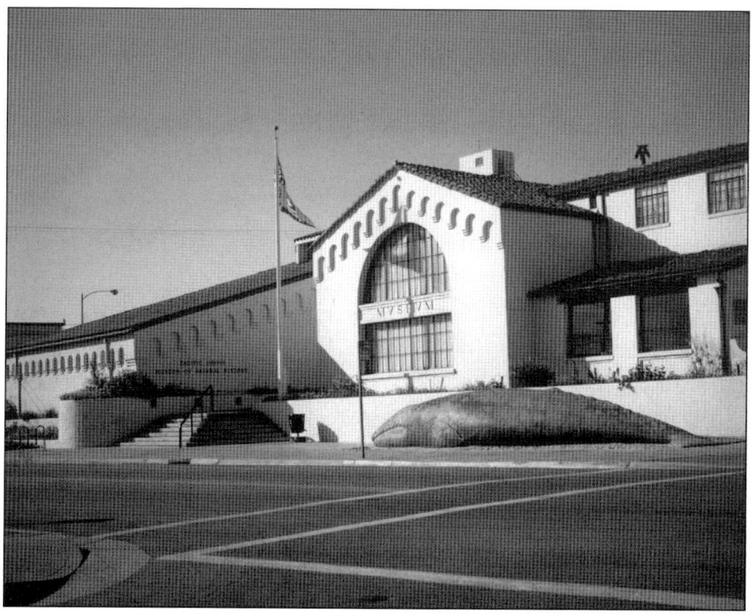

This photograph shows the museum's new addition, which more than doubled the size of the long-standing building, as it appeared in 1984. Construction on the new museum was made possible by a team of community members, who held a fundraiser to pay for the work, as well as from other sources. (Courtesy Pacific Grove Museum of Natural History.)

Bibliography

Dainton, Mary. "The Story of Sandy the Whale." Pacific Grove Museum of Natural History archives.

Heath, Dr. Harold. "Pacific Grove Museum of Natural History." Pacific Grove Museum of Natural History collection.

Horizons: a publication of Pacific Grove Museum of Natural History Association. Pacific Grove, CA: Hana Media Group, Summer 2010.

Kohrs, Donald G. *Chautauqua: The Nature Study Movement in Pacific Grove*. Donald G. Kohrs, 2015.

Miller, Evelyn Grantham, Martha Pyre Larson, and Peter Rashkin, eds. *Pacific Grove, California: The First 100 Years: A Glimpse of the Past that Shaped the City's Present*. Pacific Grove, CA: The Pacific Grove Press, 1975.

www.discoverpacificgrove.com

www.pacificgrove.org

www.pgmuseum.org

About the Organization

In 2009, the museum underwent an important change. The City of Pacific Grove and the newly formed Museum Foundation of Pacific Grove partnered to operate the long-standing institution, which meant the city retained ownership of the building and the collection. Management of the collection and building, meanwhile, became the responsibility of the Museum Foundation of Pacific Grove. Lori Mannel became the first executive director for the museum and was instrumental in the transition process. Today, Jeanette Kihs works as the museum's executive director and manages a staff of more than 12 people. The museum continues to grow, both behind the scenes and for all the world to see. Currently, educational outreach programs include the Long-term Monitoring Program and Experiential Training for Students (LiMPETS) and the Watershed Explorers program, operated in partnership with the Monterey Peninsula Regional Park District. Museum volunteers can also be found at 250 Ridge Road off Lighthouse Avenue, home of the Monarch Grove Sanctuary, during monarch butterfly season in Pacific Grove. These are but a few of the programs that have helped grow the region's natural history. Exhibits within the museum continue to evolve as well. From natural history-themed quilt shows to science illustration exhibits by California State University Monterey Bay students, the museum is often abuzz with new wonders. Special events, from weddings to birthdays, include the museum's evening Night Owl offerings and the annual Butterfly Ball, featuring fine dining and dancing. Take a moment to explore the museum's website at www.pgmuseum.org.

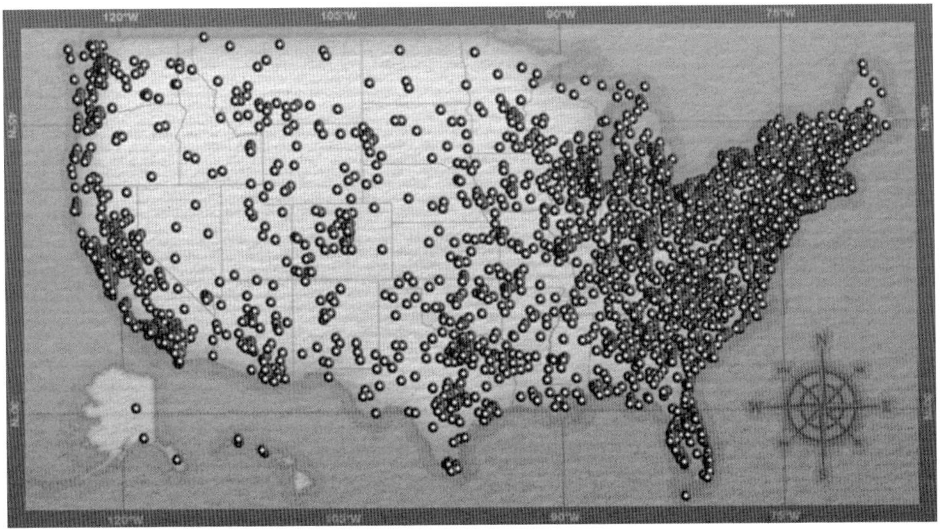